How to Build with Stone, Brick, Concrete & Tile

No. 980
$8.95

How to Build with Stone, Brick, Concrete & Tile

By Leo D. Maldon

TAB BOOKS
Blue Ridge Summit, Pa. 17214

FIRST EDITION

FIRST PRINTING— JUNE 1977

Copyright © 1977 by TAB BOOKS

Printed in the United States
of America

Hardbound Edition: International Standard Book No. 0-8306-7980-4

Paperbound Edition: International Standard Book No. 0-8306-6980-9

Library of Congress Card Number: 77-80389

Cover photo courtesy of National Concrete Masonry Association.

Contents

Acknowledgments

Thanks to the many people who helped make this book possible, particularly the Portland Cement Association, the National Concrete Masonry Association, the Brick Industry Association and the Building Stone Institute. The Portland Cement Association is thanked, especially, for its permission in using illustrations from its publications, *Cement Mason's Guide to Building Concrete Walks, Drives, Patios, and Steps*, and *Concrete Improvements Around the Home*.

Preface

There are hundreds of projects you can do in and around your home with concrete, brick, clay tile, or stone. Armed with the basic knowledge contained in these pages, the do-it-yourselfer can construct breathtaking patios, walks, retaining walls, barbecues, fireplaces, chimneys, and even a vinyl-lined swimming pool.

In fact, this book is designed to furnish the home handyman with all the information about the tools, plans, tips and tricks he needs to construct anything from functional formwork to glamour masonry. The savings in hard-earned dollars will be tremendous. The satisfaction gained from a beautiful job well done will be immeasurable.

The book begins with concrete work—the proper mix, tools, and materials. The reader will see how to get the proportions right and take the first most important step of any project: planning. Basic considerations and building codes are covered for driveways, walkways, and patios. Included are the simple but critical steps to proper grade preparation and formwork, plus instructions for mixing, placing, striking off, and floating concrete. The finishing process is also detailed—control joints, final floating, and troweling are covered. Curing is explained, along with hot and cold weather tips, and information about special finishes that can add that extra professional pizzazz.

From concrete slabs the book progresses to walls, foundations, and steps, and then on to concrete block and brick. Masonry information includes mortar, ties, tools, lintels, sills and coping, piers and pilasters. From projects like barbecues, screen walls, sandboxes, and mailbox posts, the book progresses to the more specialized techniques for arches and fireplaces. Even structural clay tile is included, along with the oldest and most rugged building material of all: stone.

And, since trees have strong roots, children have hammers, and northern climes have frosts, attention is also paid to patching and repairing any work you've done.

Let your friends gaze with awe on the professional masterpieces they will encounter at your home. This book covers all the do-it-yourselfer needs to know to construct safe, exquisite structures.

Leo D. Maldon

Part 1 : Concrete

1

The Proper Mix, Tools, and Materials

A "cement" driveway is not cement at all. If it were made of cement, you would have nothing but a dustbowl leading up to your garage. Cement is but one of the many ingredients of *concrete*. Your driveway, sidewalk, house foundation, and patio are made of concrete, not cement (Fig. 1-1).

This is not to underestimate the importance of cement. Although it makes up only seven to fourteen per cent of the content of concrete, it is the ingredient which holds the rest of it together. In a very real sense, cement is the glue. Technically, when mixed with water, it is called *cement paste*.

INGREDIENTS

The cement used in concrete is *portland* cement. It is a soft, fine, grayish-green powder manufactured from pulverized limestone and other ingredients. The term portland was applied to it by an Englishman named Joseph Aspdin in the early nineteenth century. He thought the concrete made from this cement looked a lot like *Portland stone*, a widely used building material of those times.

Aggregate

When used in construction, portland cement and water form an adhesive chemical compound which fills in the spaces

Fig. 1-1. The "cement" driveway shown is actually made of concrete, which is a mixture of cement and other ingredients.

and binds the *aggregate*. Aggregate is inert, inorganic material—like rocks and stones—which is the basis for the durability and strength of the concrete.

Modern portland cement is made from limestone, as mentioned, plus other materials such as clay, shale, and slag from blast furnaces. Its principal chemical components are lime, silica, aluminum, and iron. The materials are burned in a rotary kiln at 2700° Fahrenheit, which welds the ingredients into *clinkers*. After cooling, the clinkers are pulverized with a small amount of gypsum to regulate setting time.

The water in concrete constitutes between 15 and 20 per cent of the volume, while the aggregate is between 66 and 78 percent of the volume (Fig. 1-2). Any type of tapwater is fine for concrete work, as long as it does not have a pronounced odor.

Aggregates are divided into two sizes, fine and coarse. Fine aggregate is always made of sand, and coarse aggregate is usually made of gravel, crushed stone, or blast furnace slag.

Natural sand is the most commonly used fine aggregate; however, manufactured sand, made by crushing gravel or stone, is also available in some areas. Sand should have particles ranging in size from 1/4 inch down to the dimensions of a particle of dust. Mortar sand should not be used for making concrete since it contains only the small particles.

Gravel and crushed stone are the most commonly used coarse aggregates. They should consist of particles that are sound, hard, and durable, not soft or flaky, and should have a minimum of long, sliver-like pieces. Particles range in size from 1/4-inch up to a maximum of 1 1/2 inches. The most economical concrete mix is obtained by using coarse aggregate that contains particles of the largest practical size. In slabs for driveways, sidewalks, and patios, for example, the largest practical particles would be about one-third the slab thickness. Accordingly, four-inch slabs should use coarse aggregate with particles that range to a maximum size of one inch, while slabs of five- to six-inch thickness should use aggregate that has particles of a maximum 1 1/2 inches. Use aggregate with maximum particles of one inch for steps. All sizes may not be available locally; but, within the above limitations, try to use aggregate with the largest particles readily available. All aggregates must be free of excessive dirt, clay, silt, coal, or other organic matter such as leaves and roots. These elements prevent the cement from properly binding the aggregate particles, and result in porous concrete with low strength and durability.

If you suspect that the sand contains too much extremely fine material such as clay and silt, check its suitability for concrete by conducting a *silt test*. Fill an ordinary quart

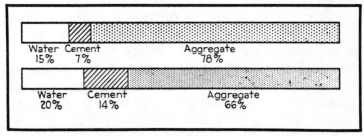

Fig. 1-2. Graphs show the range of ingredients possible for acceptable concrete mixes. Top bar is a lead mix of a stiff consistency, using large aggregates. Bottom represents a rich, wet mix using small aggregates.

Fig. 1-3. An example of well-graded aggregates. The particles range in size from 1/4 inch to 1 1/2 inches.

canning jar or milk bottle to a depth of two inches with the sand. Samples should be taken from at least five different locations in the sand pile and should be thoroughly mixed together. Add clean water to the sand in the jar or bottle until it is about three-quarters full. Shake the container vigorously for about a minute. Use the last few shakes to level off the sand. Allow the container to stand for an hour. Any clay and silt present will settle out in a layer above the sand. If this layer is more than three-sixteenths inch thick, the sand is not satisfactory. The clay and silt should be removed by washing.

Good concrete aggregates should have a range of particle sizes but not too much of any one size (Fig. 1-3). The big particles fill out the bulk of a concrete mix and the smaller ones fill in the spaces between the larger ones. Aggregates with an even distribution of particle sizes are said to be *well graded* and produce the most economical and workable concrete. Don't use mixtures of fine and coarse aggregates taken directly from gravel banks or stone crushers. These aggregates usually contain an excess of sand and too little coarse material.

Air-Entraining Agent

Another ingredient of most concrete mixes is a chemical *air-entraining agent*. This material is important if the concrete will be exposed to alternate cycles of freezing and thawing. Though it may not be necessary for use in the Deep South, it should always be used for outdoor installations in the North, and is advised in any climate where there may be freezing weather for even a few days in the winter.

Hardened concrete always contains some minute particles of water. When the water freezes, it expands, causing pressure that makes *scales* on the concrete surface. The air-entraining agent causes microscopic bubbles of air to form inside the concrete, acting as relief valves for the expanding water. Air-entrainment also helps concrete to resist the effects of salt deicers. What's more, the tiny air bubbles act as ball bearings in the mix, making the mixing process easier.

There are several brands of air-entraining agents. Although the proportion of agents in the mix usually averages around six percent of the volume, check with your dealer and manufacturers' directions for the exact percentage of this material that should be added. If you do use it, be sure to add it in a portable mixer or order it with ready mixed concrete. Hand mixing does not do an adequate job of dispersing it.

GETTING THE RIGHT PROPORTIONS

It is important in concrete work to achieve the right proportions of cement, water, and aggregate. If one small job is all you are working on, a bag or two of premixed concrete such as Sakrete is all you need (Fig. 1-4). Just add water according to the directions on the package. For large jobs, you can order readymix from a local concrete service (see the Yellow Pages of the phonebook). The "cement" truck will not only bring a perfect mix, but will save many hours of back-breaking hand mixing. (See Chapter 4 for details on how to order readymix.)

It is less expensive, of course, to mix your own concrete. Sometimes you will find that a concrete truck cannot get to the site, and you will have to mix your own anyway. And, unless you have several willing helpers, you will not be able to handle

Fig. 1-4. For small jobs, a bag of premixed concrete is easy to use. It contains exactly the right proportions of ingredients.

the large loads that readymix suppliers deliver. For one or two workers, it is easier to mix smaller loads, divide the project into stages, and place the concrete loads in one stage at a time.

Portland cement is purchased in bags from local building and mason supply houses. Bags weigh 94 pounds in the United States and are one cubic foot in volume. In Canada, bags weigh 80 pounds each and hold about 7/8 of a cubic foot.

For most applications of concrete, the following proportions are ideal:

   94 pounds of cement (one U. S. bag)
 215 pounds of sand
 295 pounds of coarse aggregate
    5 gallons of water

A regular bathroom scale is accurate enough for weighing the materials. Use a large, three- to five-gallon galvanized bucket to hold the material, but be sure to weigh the pail first and deduct its weight from the weight of the material. (Setting the scale at zero when the empty bucket is on it may be easier.) Don't put any more in the bucket than you can handle. Put the sand and aggregate into three or four buckets of equal weight. Once you get the right weight established, mark a line on the bucket for each ingredient and put the scale back in the bathroom. See Chapter 4 for details on exact proportions using the various-sized aggregates.

A simpler, but less accurate, guide is based on volume. To accomplish the same average mix as above, use one part portland cement to 2 1/4 parts sand and three parts aggregate. Add five gallons of water for each bag. These one-bag formulas should yield approximately 1/6 cubic yard of concrete. See Chapter 4 for other proportions.

The proportions above are for typical wet sand. Sometimes moisture in the sand or the size of the aggregate will throw off the above proportions, so be sure to make a test batch before forging ahead with the whole job. If the test batch is too wet, add about five to ten percent more sand and aggregate, and test again. If the mix looks too stiff, add proportionally more water *and* cement (never water alone) to the test batch. In subsequent batches, decrease the amount of sand and aggregate to accomplish the same purpose.

A mix that is too sandy requires a shifting of the aggregate proportions. Add about four pounds of stone to the test batch. In subsequent batches, decrease the sand by about two pounds and add two pounds of coarse aggregate. If the mix is too stony, change the aggregate ratios by adding sand and subtracting stone in future batches. To save the test batch, add about four pounds of sand.

REQUIRED TOOLS AND MATERIALS

The most important tool in working with concrete is your back. Placing and working concrete is demanding physical work, so don't attempt it if you are allergic to hard labor. Also, stay away from concrete work if your body is getting more than a little flabby, or if you have a serious medical problem like a heart condition.

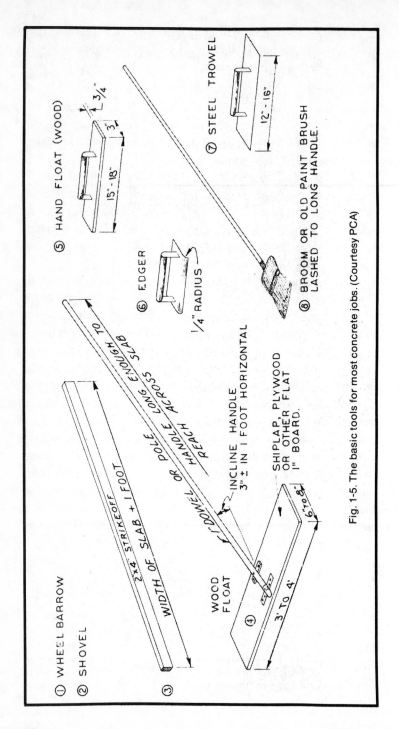

Fig. 1-5. The basic tools for most concrete jobs. (Courtesy PCA)

① WHEEL BARROW

② SHOVEL

③ 2"×4" STRIKEOFF SLAB + 1 FOOT
WIDTH OF SLAB
1" DOWEL OR POLE LONG ENOUGH TO REACH ACROSS HANDLE ACROSS SLAB

④ WOOD FLOAT
INCLINE HANDLE 3" ± IN 1 FOOT HORIZONTAL
SHIPLAP, PLYWOOD OR OTHER FLAT 1" BOARD.
6" to 8"
3' TO 4'

⑤ HAND FLOAT (WOOD)
3/4"
3"
15'–18"

⑥ EDGER
1/4" RADIUS

⑦ STEEL TROWEL
12"–16"

⑧ BROOM OR OLD PAINT BRUSH LASHED TO LONG HANDLE.

The other tools you will need depend on how much of this work you plan on doing. If a small slab or some patchwork is all you have in mind, a piece of 2 × 4 and a pointed trowel will be about all you will need. Jobs of larger size, however, will also require the use of the tools shown in Fig. 1-5 and some not shown and listed below:

wood hand float—used for compacting and for rough finishing. Often used as the final finish for non-slip texture.

steel rectangular trowel—for producing a dense, smooth final finish.

long-handled bull float—same as a hand float, but reaches areas too far away for a hand float.

darby—used instead of a 2 × 4 for striking off the concrete, and also for initial smoothing instead of the float.

edger—makes a nice, smooth edge on concrete surfaces.

groover—for making control joints.

short-handled, square-end shovel—for distributing concrete throughout the forms. *Not* a garden spade (although a garden spade comes in handy for digging and removing sod).

buckets—for accurate measuring.

concrete rake—also known as a *mortar hoe*; has one or two large holes in it for easier mixing and pushing concrete into place. A garden hoe will do in a pinch. (Fig. 1-6).

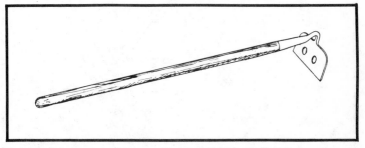

Fig. 1-6. A concrete rake or mortar hoe has holes in the blade and works better than a garden hoe (though in a pinch, a garden hoe will do the job all right).

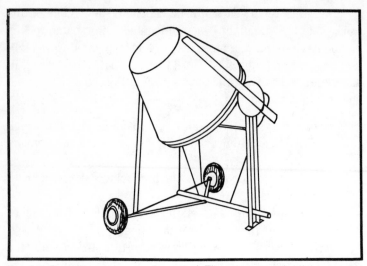

Fig. 1-7. Portable mixers may be rented. They are a must for any extensive home-mixed project.

portable mixer—a must for jobs of any size. Can be rented or bought used (Fig. 1-7).

kneeboards—these are made on the job, and are used to kneel on while working on the concrete itself in order to distribute your weight. Made of a piece of plywood, about 2 1/2 × 2 1/2 feet, with 2 × 2 or 1 × 3 edges to keep from slipping off (Fig. 1-8).

mortar box—for smaller jobs and hand mixing.

wooden chutes—also made on the job out of one-inch lumber, for pouring concrete in hard-to-reach places.

Fig. 1-8. Kneeboards can be made from plywood and lumber scraps.

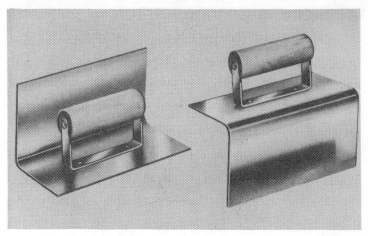

Fig. 1-9. Inside and outside edgers for step work.

wheelbarrow—for getting the mix to the job when the mixer cannot get close enough. Rubber-tired, heavy-duty types are best.

brooms and brushes—for making non-slip surfaces. Notched trowels and other tools can be used for other surfaces (see Chapter 7).

step edgers—for making edges of steps. Use depends on how steps are formed (detailed in Chapter 8). (Fig. 1-9).

garden hose—for your water supply.

burlap or *sheet plastic*—for curing the concrete after it is finished.

dimension lumber—nominal one- and two-inch lumber in various sizes for making forms.

nails—for making forms. Since the nails must be removed, the double-headed type is the best.

reinforcing bars—metal bars used to strengthen concrete in certain applications.

There are several professional machines that can be used for specialized applications. The uses of these and the others above will be explained in detail as they fall into sequence.

2

Planning—The Important First Step

Proper planning is the important first step for any project, and even more so with concrete work. Mistakes in this medium cannot be covered up with molding or cut off with a chisel. Once concrete sets, it's a herculean job to remove it. It's also poor practice to add more concrete after pouring. So, make sure that you do it right from the beginning.

A typical *slab* is shown in Fig. 2-1. If you plan your project right, and proceed with the other steps in this book, you should not have any difficulty in laying down a slab. A slab is a flat section of concrete such as a sidewalk, driveway, or patio. Steps, walls, and castings are more difficult, and should be attempted after you have had some experience with simpler work like a slab. Of course, if you do not *need* a driveway, walk, or patio, and *do* need some steps, go ahead and try it, but you should read all the chapters in this section first. Then proceed to Chapter 8.

SOME BASIC CONSIDERATIONS

When planning, it is important to realize that walkways, drives, and patios are part of the landscape. Professional mason contractors are not always interested in aesthetics, and have been known to take the path of least resistance and

Fig. 2-1. Any home project should be planned to blend in with the rest of its surroundings. This patio fits perfectly with the house, fence, shrubbery, benches.

highest profit. The do-it-yourselfer can afford to take the time and initiative to make his project blend in as much as possible with its surroundings, avoiding being intrusive. Look at all the possibilities before plunging ahead.

Do not, for example, automatically run your driveway in a straight line from the garage to the street. At least consider the possibility of making a curved, graceful drive. The same goes for sidewalks and patios. There is a limit, of course. No one wants a corkscrew walk that takes twice as long to get where it is supposed to go than a straight one.

But do consider the effect of your work on the eye, in addition to the project's practicality. A curving walk or driveway, as shown in Fig. 2-2, will take a little extra work in forming and placing, plus more money for forms and materials; but it may be a lot more pleasing to look at. The important factor is that it blends in nicely with the rest of your property.

The topography of your lot should play an important part in your decision. A flat lot may do just as well with straight lines; but on a hilly plot, the lines may be more practical and pleasing when the drive and walkway follow the contours of the land.

There are other factors, too. How many *steps* will there be if you build a straight walk? Steps are more difficult to build than slabs. On the other hand, you do not want to build a walk with too steep a grade, especially in snowy areas. Consider the use of precast rounds, as illustrated in Fig. 2-3, or other castings. This type of planning is highly individualized. Don't automatically think in terms of squares and rectangles. An egg-shaped patio, like that shown in Fig. 2-4, can be very

Fig. 2-2. Straight, plain driveways have their place, but look how much more attractive this curving drive is. Exposed aggregate and redwood divider strips further enhance the effect.

Fig. 2-3. Precast, exposed-aggregate rounds give this hilly homesite an easy and attractive entranceway.

pleasing to the eye. There are certain standards for grades (see specific applications below), but sometimes they can be hard to apply to specific plots. For really difficult terrain, a landscape architect can be a big—though expensive—help.

Drainage is another vital consideration. It is amazing how many handymen and builders construct driveways that drain into the garage or into areas that are already subject to erosion. Flat, impermeable slabs—driveways in particular—act like conduits for heavy rains. Your neighbor will not be happy if these slabs drain onto his property, and you will not be very delighted either if you have a garage full of water

Fig. 2-4. A patio doesn't have to be rectangular.

after a downpour. Always make sure that the water is properly directed so it will not damage your lawn or flower bed either.

CHECK BUILDING CODES

Before beginning *any* type of construction, inside or out, it pays to check the building codes in your area. Most communities require a building permit, before you begin, to insure that the work is done according to code. These statutes vary considerably, so ask at the city, town, or county planning or building commission. Permits are almost always needed for driveways and sidewalks that cross the *public way*, which is usually that area between the front of your property and the street. On city sidewalks that run along the street, grades, concrete composition, and other factors may be strictly regulated.

PLANNING DRIVEWAYS

The time to plan and build your concrete project is early in the construction season before the hot days of summer. There will be less chance of premature drying out from hot sun and winds at this time.

Driveways for single-car garages or carports are usually 10 to 14 feet wide, with a 14-foot minimum width for curving drives (Fig. 2-5). In any case, a driveway should be at least three feet wider than the widest vehicle it will serve.

Long driveway approaches to two-car garages may be single-car width, but must be widened near the garage to provide access to both stalls. Short driveways for two-car garages should be about 16 to 24 feet wide all the way from the street to the garage.

How thick a driveway should be depends primarily upon the weight of the vehicles that use it. For passenger cars, four inches is sufficient, but five or six inches is the recommended thickness if an occassional heavy vehicle, like an oil truck, will use it.

If the garage is considerably above or below street level and is located near the street, the driveway grade may be critical. A grade of 14 percent (1 3/4 inches vertical rise for each running foot) is the maximum grade recommended by the Portland Cement Association. The change in grade should be gradual enough to avoid the scraping of the car's bumper or underside. The most critical point occurs when the rear wheels are in the gutter as a vehicle approaches from the street (Fig. 2-6).

The driveway should be built with a slight slope so that it will drain quickly after a rain or washing. A slope of 1/4 inch per running foot is recommended in the direction of drain. The specific direction will depend on local conditions, but usually it should be toward the street. A crown or cross-slope may be used for drainage where the drive is sloped toward the garage;

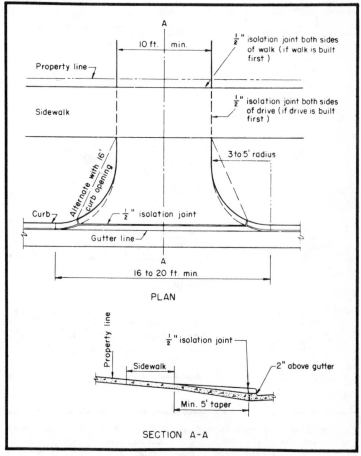

Fig. 2-5. Details of a typical driveway plan.

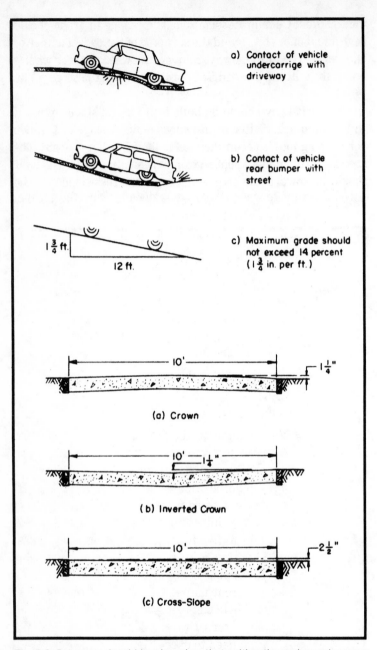

a) Contact of vehicle undercarrige with driveway

b) Contact of vehicle rear bumper with street

c) Maximum grade should not exceed 14 percent ($1\frac{3}{4}$ in. per ft.)

$1\frac{3}{4}$ ft.

12 ft.

10'

$1\frac{1}{4}$"

(a) Crown

10' $1\frac{1}{4}$"

(b) Inverted Crown

10'

$2\frac{1}{2}$"

(c) Cross-Slope

Fig. 2-6. Driveway should be sloped so that neither the undercarriage nor the rear bumper of your car will make contact. Maximum grade is 14 per cent (top drawing.) Drainage can be provided by any of the methods illustrated in lower drawing.

a dry well of adequate size should be built in front of the garage.

The part of driveway between the street and public sidewalk is usually controlled by the local municipality. The municipality should be consulted when a driveway is built after the street, curbs, and public walks are in place. If curbs and gutters have not been installed, it is advisable to end the driveway temporarily at the public sidewalk or property line. An entry of gravel or crushed stone can be used until curbs and gutters are built. At that time, the drive entrance can be completed to meet local requirements.

If the driveway is built before the public walk is completed, it should meet the proposed sidewalk grade and drop to meet the gutter (if no curb is planned) or the top of any low curb.

While in the planning stage, consideration should be given to other elements that can make a driveway a beautiful approach to a home rather than just a pathway to the garage. Consider using exposed aggregate or other fancy concretes (see Chapter 7 for details) for offstreet parking, turnaround areas for safe head-on entry to the street, or multi-use paved areas for games.

PLANNING WALKWAYS

Private walks leading to the front entrance of a home should be three to four feet wide. Service walks connecting to the back entrance may be two to three feet wide.

Public walks should be wide enough to allow two people walking abreast to pass a third person without crowding. The width will vary with the amount of pedestrian traffic on your street. A width of four to five feet is advisable for quiet areas with single-family housing. A greater width is required near churches, schools, shopping centers, and other areas where walks are used by a great number of people. Walks serving apartment dwellers should be at least eight feet wide. Those in commercial shopping areas should be 12 feet wide. On the other hand, there is no need to exceed four feet in width along rural highways (if you need a sidewalk at all).

In areas where there is a frequent discharge of automobile passengers (such as homes near churches or restaurants), it

Fig. 2-7. Residential sidewalks should be designed for safe bike-riding, walking, skate-boarding.

may be wise to install a courtesy walk next to the curb. This is not only to accommodate the people, but to save your grass. Such walks are usually 18 to 30 inches in width.

In some neighborhoods, all sidewalks are built adjacent to the street curb. This allows larger lawns and proper footing for people alighting from vehicles. Such sidewalks, however, give pedestrians and cyclists less protection from street traffic than walks set back from the curb (Fig. 2-7). A study of pedestrian and bicycle accidents in a surburban area found that the accident rate with five-foot walks next to the curbs was 2 1/2 times the rate for areas with similar walks set back 7 to 14 feet.

Private residential walks should not be less than four inches thick. Walks in commercial or business areas are generally five to six inches thick. Your local building code may specify width, depth, and other aspects of public sidewalks.

It is customary to slope walks 1/4 inch per foot of width for drainage. Where walks are next to curbs or buildings, the slope should be toward the curb and away from the building over the full width of the walk. In some areas where side drainage permits, walks built with a crown or slope from center to edge are desirable. Certain conditions may require that a slope other than 1/4 inch per foot may be used—for example, where a new walk meets an existing driveway or alley. In such cases, the cross-slope of the walk may be increased to 1/2 inch per foot.

For the convenience of pedestrians, sidewalk approaches at street intersections can be planned to eliminate steps by providing a gentle ramp from walk to street. This is ideal for cyclists, roller skaters, housewives with baby buggies, and elderly pedestrians.

Sidewalk design is closely linked with street, pavement, curb, and gutter design. If possible it is best to construct them at the same time. By coordinating sidewalks with streets, curbs, and gutters, a pleasing effect can be obtained (Fig. 2-8).

Fig. 2-8. Commercial sidewalks must be wide enough to handle heavy traffic, but there is no need to make them dull. This attractive, precast concrete walk adds to a graceful scene.

PLANNING PATIOS

Patio planning takes a good deal of thought, with family convenience being the primary consideration. In most cases, this means that the patio should be located directly outside the kitchen or family room, preferably linked with sliding doors. This is a particularly good idea if you entertain frequently. But this location of a patio should not be determined before looking at other possibilities and discussing the matter with the others in your family.

Another possible site, perhaps even better for frequent entertainers, is off the living room or the dining room. Some couples may prefer a small, secluded spot outside the master bedroom or even the bath—great for sunbathing. Of course, there is no reason why you cannot have more than one patio. There could be a large, family patio near the kitchen plus a smaller more private, fenced-in patio off the master bedroom.

You may not even want the patio to be attached to the house. If there is a treed area, or a spot with a view on the far corner of your lot, this may be the best location of all, though it may not be as convenient. What about the weather? The winds should be taken into consideration. In areas where the weather is hot, take advantage of the prevailing winds. In colder areas, you may wish to shield the patio from the breezes.

The *purpose* should also be considered. If outdoor dining is your weakness, the patio should be close to the kitchen and a fuel supply for the barbecue. For sunbathing, the privacy of the patio may be the main consideration.

Sunlight may be the most important aspect of all. Depending on where you live, the patio should be oriented to either take advantage of, or block out, the sun. A patio that faces south will almost surely need a roof or some other sort of sunscreen. An eastward-facing patio gets the morning sun, while a westward-facing patio will be exposed to the hotter midday sun. An attached patio that faces north will be shaded by the house most of the day. This can be good or bad, depending on the prevailing weather conditions and how you think you will use the patio.

The patio is an extension of the home, although not necessarily a physical part of it. It should be planned to

Fig. 2-9. Patios should be planned for comfort, convenience, and view. Most are located next to the house, but they don't have to be.

Fig. 2-10. Another example of an attractive and convenient patio.

complement the architecture of the house, as do the types shown in Figs. 2-9 and 2-10. Shape and size must be considered in relation to the house and to the size and shape of the lot. In most cases, the patio should be at least as large as the main rooms of the house, incorporating lots of room for oversized lounging furniture. The more intimate, off-the-bedroom type patio might be smaller and surrounded by fences or shrubbery.

All concrete patios should be pitched slightly to avoid rainwater collecting on the surface. About 1/4 inch per foot should be sufficient. The slope should, of course, be away from the house if the patio is attached to it. Do not, by the way, opt for a concrete patio without considering the other types. A brick-in-sand patio, for example, blends in more nicely with certain types of architecture, and does not need to be pitched because water will drain between the bricks. Usually, brick will look better with a brick house, stone with a stone house. See the appropriate chapters on how to build other types of patios. A concrete patio goes with any style house, however, and is particularly effective next to a pool. Poolside patios should be planned to have a non-skid surface (see Chapter 7).

3

Grade Preparation
and Formwork

Concrete work can be compared to baking a cake. Getting the
proper ingredients together in the right proportion is very
important, but you also have to find the right pan and prepare
it. The "pan" in this case is the framework. Instead of
greasing the pan, we have to prepare the ground. No matter
how well you mix the ingredients, the "cake" won't come out
right in a defective pan. You can't make a sponge cake in a
flat, layer-cake form.

SOIL PREPARATION

After poor mixing, the next most frequent error in concrete work is poor soil preparation. Slabs will settle, crack,
and fall apart in poorly prepared or compacted soil. The
subgrade should be free of all organic matter such as sod,
grass, roots, soft and mucky ground. Dig out all such material,
as shown in Fig. 3-1. If some spots are very hard and others
very soft, the concrete will surely crack from settling in the
softer areas. Break up the hard areas and disperse the soil.
The subgrade must provide uniform support for the concrete.

On the other hand, do not routinely remove whatever soil
is under your slab and replace it with fill. If the soil is already
reasonably uniform and free from vegetable matter, it is

41

Fig. 3-1. Poor soil preparation is one of the most common errors in concrete work. All organic materials must be removed as shown.

better left alone. Nature has already done your compacting for you. Most sandy soils merely need tamping of the top portion which was disturbed by your spade. Undisturbed soil is better than soil that has been dug out, replaced, and poorly compacted.

If the soil is soft and organic, it should be dug out to a depth of four to six inches below the concrete bed (or about 8-12 inches in all, depending on the thickness of your slab). If the soil is exceptionally mucky, you may have to dig deeper, but do not fill in the entire section at once. Fills should be made in four-inch sections, compacted, and followed by another layer. The fill should be extended at least a foot in all directions around the slab to prevent undercutting during heavy rains.

Common fill materials are sand, gravel, crushed stone, or blast-furnace slag. Sand is preferred because you can level it more easily. You can also use leftover soil from high spots as long as it is of the same quality as the rest of your subgrade. The latter is probably the simplest method if you are just filling in a few low spots. Compact this fill material completely, however.

After the fill material has been graded, it must be compacted with a hand tamper for small jobs (Fig. 3-2) or preferably for large jobs, with a roller or vibratory compactor (Fig. 3-3). You should be able to get these machines at a local

Fig. 3-2. Fill materials must be tamped down thoroughly.

43

Fig. 3-3. For large jobs, a roller can be rented to do a faster and better job of fill compaction.

rental supplier. Before placing the concrete, the subgrade is dampened with a garden hose. Otherwise, the ground will absorb the water from the concrete. Do not leave standing water, however, or any soft or muddy spots.

When building walls, steps, or other structures that are deep and require a lot of concrete, you can fill the cavity with any kind of inorganic material such as old concrete, rocks, or metal (Fig. 3-4). Sand is then added, filling all the holes and cracks completely. Be sure to compact the sand thoroughly on top and around this type of fill. Leave at least 4 to 6 inches all around for fresh concrete.

SETTING FORMS

Steel stakes and forms are often used by professional masons, but wooden forms are perfectly acceptable. Nominal two-inch lumber is used in most form-building, although one-inch boards and thin plywood are usually used for making curves. For walls and steps, 3/4-inch plywood is preferred,

Fig. 3-4. A new set of steps can require a lot of new concrete. The greater part of the cavity can be filled with inorganic materials like old, broken concrete.

Fig. 3-5. A line level may be used for small jobs, but large projects are best graded with a builder's level.

although several 2 × 8s or 2 × 10s can be placed on top of each other.

It is important in most slabs to establish the proper grade. Small jobs, and complex slopes are sometimes best determined with your own eye, but for any larger slabs you'll have to use a builder's level or line level to establish the grade (Fig. 3-5). Stakes should be driven into the ground around the perimeter of the slab, and marked with a builder's level. One person should look through the level while the other works a measuring rod or rule that has been set on an established grade. The rule is placed against a stake and moved up and down until the desired grade is read through the level. Put a mark on the stake at the bottom of the rule, then attach a string line tightly to the stake at this mark. Using the level, mark the other stakes and string the line tightly from stake to stake. The other stakes are set in line with the string.

A line will not give an exact setting, but is usually all right for do-it-yourself work. Drive stakes in at the corners of the slab, and in between the corners, about every eight feet. The stakes at the low ends should be high ones, for reasons soon seen. Run a line from one corner to the next, as shown in Fig. 3-6, and place the line level in the center. If the distance is a long one, use some of the intermediate stakes, and repeat the procedure as you go along. The line must be very taut (Fig. 3-7). Make a mark where the line is level, then measure down to where the top of the form should be. If the grade, for example, is 1/4 inch per running foot, and the distance between stakes is eight feet, the drop should be two inches. Mark the stake two inches below the top of the line. That is where the top of the form should be. As you can see, the mark can go several inches below the top of the line at the low end, which is why high stakes were recommended for low corners.

The size of the forms and stakes depends on how deep the excavation is and how much concrete pressure will be exerted. For most sidewalks, patios, and other slabs, 2 × 4s are best.

Fig. 3-6. Once grade marks have been established, tie a line tightly to a corner stake, run it to the next corner, and attach a line level to the center.

Fig. 3-7. Forms should be nailed to stakes with the top inner edge flush with the line.

Six inch thick slabs or larger will take 2 × 6 forms. You can use 2 × 4 forms for five-inch slabs, but 2 × 6 is preferable.

Remember that these lumber sizes are strictly *nominal*. The term "2 × 4" represents the size, in inches, before the wood is shrunk, planed, and surfaced. For the past several years, dimension lumber has been, in reality, a half-inch less than the nominal size. A 2 × 4, in other words, is actually 1 1/2 inches by 3 1/2 inches. Any new lumber you buy will follow that formula, but if you use old lumber, it could be only 3/8 inches less, instead of 1/2 (2" × 4" = 1 5/8" × 3 5/8"). This is not crucial, but it should be taken into consideration. The slab dimension should be the true size, not the nominal size, and therefore the bottom of your forms will be either 1/2 or 3/8 inches above the bottom of your excavation.

As mentioned above, one-inch lumber or 1/4-inch exterior plywood is used for curves. One-inch lumber can also be used for straight runs, but it requires more staking to prevent bulging. For most slabs, 2 × 2 staking is sufficient, but 2 × 4 or even larger stakes are necessary for walls, steps, or other applications with lots of concrete pressure. Stakes should be driven every four feet for two-inch formwork and every two to three feet if using one-inch lumber. On gentle curves, regular spacing is sufficient, but stakes must be driven every one to two feet on short-radius curves.

For ease in placing and finishing concrete, drive all stakes slightly below the top of the forms. Wood stakes can be sawed off flush, if that is easier. Drive all stakes straight and true to insure plumb. If there is any doubt that the stake will hold, it can be braced by driving another stake diagonally and nailing the two together (Fig. 3-8). Bracing and staking are more complex for steps (see Chapter 8).

You nail all stakes into the forms by holding the form boards with your foot (Fig. 3-9). Since this can be difficult, especially when nailing into thinner lumber, set temporary stakes inside the forms to hold the forms, then remove them after everything is nailed up. Double-headed nails are recommended for all formwork, since removal is easier and cleaner.

Fig. 3-8. For thick slabs, or whenever you have a doubt about whether a stake will hold, drive in a brace on the diagonal and nail it to the stake.

When making curves with plywood, the outside grain should be vertical to permit easy bending. For extra strength on curves, 2 × 4s can be used by wetting and gently bending them. For a shorter radius, make saw kerfs (notches) on the inside of forms and bend them so that the cut is close. For medium-radius curves, one-inch lumber is your best bet (Fig. 3-10).

Forms do not necessarily have to be made for removal. They may be left in for decorative purposes. Divider strips of wood may also be left in place. These strips eliminate the need for control joints. Although you do save on some removal and jointing work by leaving these forms, the primary use for this type of forming is decorative and it is usually reserved for

Fig. 3-9. Use double-headed nails for easy removal. When nailing stakes to a form, brace the form with your foot.

Fig. 3-10. Curves are formed with either one-inch lumber or 1/4-inch exterior plywood, bent to the required shape and staked as shown.

Fig. 3-11. Cover wood divider strips with masking tape to protect them from abrasion and staining.

patios. There is no reason why it can't be done for walks and drives too, however.

The work you save in form removal and grooving control joints, however, you pay for in other labor. Removal forms only have to do the job and not look good, whereas the permanent forms should be carefully mitered at corners and neatly butted at the joints. Masking tape should be applied to the tops of the wood to protect them from abrasion and concrete staining (Fig. 3-11). Also, rot-resistant, more expensive wood such as redwood, red cedar, or cypress must be used.

Stakes must either be removed or, better, cut off two inches below the surface of the concrete. Galvanized nails

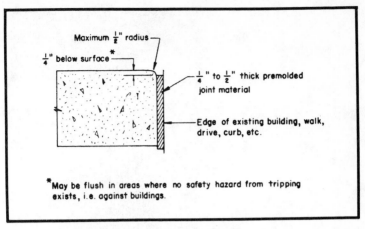

Fig. 3-12. How to build an isolation joint.

should be driven through alternating sides of dividing strips at 16-inch intervals before pouring the concrete. The same should be done for the forms, except that all nails should face inward. These nails anchor the wood permanently to the concrete. All nail heads should be driven flush with the forms, never through the top, and should be positioned approximately halfway between the bottom and the top of the wood.

After all the forms are in place, backfill under the outside to prevent the concrete from escaping. Install isolation joints at intersections such as where the patio meets the house, the driveway meets the walk, around poles, around fireplugs, or other rigid points of potential stress. Use 1/4 or 1/2 inch premolded fiber joint material available at concrete dealers. The joint material should be flush with, or 1/4 inch below, the concrete surface (Fig. 3-12).

As a final check before pouring, inspect all forms for plumb and trueness to grade. Make sure the proper slopes are allowed for drainage. Use a string line or template to insure that the subgrade is smooth and allows correct slab thickness. Dampen the subgrade and forms to prevent water erosion from the mix. Apply motor oil to the forms on small jobs to allow easier form removal. On large jobs, rent spray equipment and use professional form release agents.

4

Ordering and
Placing Concrete

For large jobs requiring a cubic yard of concrete or more, it is best to order concrete from a readymix dealer. He will put in the right proportions of the ingredients and deliver it to the site, all set to pour (Fig. 4-1). This may cost a little more than mixing it yourself would cost, but the savings in time and effort are well worth it. The readymix dealer should know exactly what to put into the mix if you explain the type of job to him, but there are certain specifications which you should make sure he follows. His mix will be more precise than yours, so take advantage of it.

Tell the dealer you want at least six sacks of cement per cubic yard, and five to seven per cent air-entrainment (unless you don't need air-entrainment, as discussed in Chapter 1). If your project is the usual four-inch slab, specify coarse aggregate with a one-inch maximum. (Table 4-1 provides proportions for five- and six-inch slabs as well.) Also tell the dealer you want a maximum *slump* of four inches. (This is a measure of workability obtained by carefully packing a conical container with concrete, overturning the container on level ground, and measuring the height of the concrete cone after a predetermined length of time has elapsed.) Also tell the dealer you want 28-day compressive strength of 3500 pounds

Fig. 4-1. The easiest way to pour concrete is to order it from a readymix dealer, who will wheel it right up to the forms wherever possible.

per square inch (this is a measure of strength under pressure after curing).

DETERMINING QUANTITY

Concrete is ordered in cubic yards, and dealers have certain minimums, usually at least one yard. To estimate how much concrete you will need, measure the area in square feet, then multiply by the depth. The math, which is relatively easy for slabs, is much more difficult for steps and other odd-shaped jobs, particularly when you have used broken concrete or other materials for fill as discussed in the previous chapter. In more difficult cases, get an estimate from a dealer or contractor.

The easiest way for most people to compute the volume of a typical four-inch slab is to multiply the length times the width in feet, then take 1/3 of that total (four inches, the common depth, equal 1/3 of a foot). Since there are 27 cubic feet in one yard, divide the last figure obtained by 27. The result is your volume in cubic yards.

To illustrate: You are building a 20- by 30-foot patio with a four-inch depth. The area is 600 square feet. Multiply this figure by 1/3, to get a volume of 200 cubic feet. Two hundred cubic feet divided by 27 gives you almost exactly 7.3 cubic yards. Allowing for possible waste and spillage, you should order eight yards. In this example we used relatively easy figures. For a five-inch slab, the math can be a little dizzying. A simpler method for most jobs is to use the following chart:

Estimating cubic yards of concrete for slabs.*

THICKNESS	AREA IN SQUARE FEET (WIDTH × LENGTH)					
	10	25	50	100	200	300
4″	0.12	0.31	0.62	1.23	2.47	3.70
5″	0.15	0.39	0.77	1.54	3.09	4.63
6″	0.19	0.46	0.93	1.85	3.70	5.56

*Does not allow for losses due to uneven subgrade, spillage, etc. Add 5 to 10 percent for such contingencies.

For a 6- by 20-foot walk, for example, multiply the dimensions to get 120 square feet. Using the chart, enter 1.23 for 100 square feet, then divide by four and add .31 for 25 feet (the nearest amount to 20), and you get a sum of 1.58 cubic yards. Since your total reflects a 125-foot area instead of the actual 120, you might possibly get away with exactly 1.5 cubic yards of concrete, but this latter figure assumes that you have graded and measured exactly, and that you won't waste a drop of concrete. It is much better to have a little more than you may need. For readymix, get whatever higher amount the dealer will sell you (probably two yards).

Readymix should be ordered at least one day ahead. Be sure to tell the dealer exactly when and where to deliver it. Make certain that everything is ready before the load arrives. Get all the tools you will need on hand, and see to it that at least one other person is around to give you a hand. If the load is a large one, three or more willing workers are recommended. If there is a lot of wheelbarrowing to be done, add one more pair of hands.

The same cautions, of course, apply to mixing your own concrete. Timing is of less importance, but you should still make sure that everything is ready before you start your mixing. You don't want to have a big batch ready to pour, then find that you have to run down to the hardware store for a trowel, shovel, or float—especially on a Sunday, when the store is closed.

MACHINE MIXING

Before doing any kind of mixing on your own, be sure to read Chapter 1 thoroughly so that you know how to achieve the right proportions in the mix. The next step is doing the actual mixing, which is best performed in a rented mixer, powered by either gasoline or electricity. An electric model is quieter and easier to operate, but you need an available power source.

Mixer sizes go by the maximum efficient concrete batch in cubic feet, usually 60 percent of the total volume of the mixer drum. The maximum batch size is usually shown on an identification plate. For proper mixing, never load a mixer beyond its maximum batch capacity. The choice of mixer size will depend on the extent of your project and the amount of concrete that you want to handle in any one batch. To mix a one-cubic-foot batch of concrete you will have to handle 140 to 150 pounds of materials.

Ingredients are loaded in the following sequence: Add all the coarse aggregate and half the mixing water before starting the mixer. If an air-entraining agent is used, mix it with this part of the mixing water.

Start the mixer, then add sand, cement, and the remaining water.

When all ingredients are in the mixer, continue mixing for at least three minutes, or until all materials are thoroughly mixed and the concrete has a uniform color.

Concrete should be placed in the forms as soon as possible after mixing. If the concrete is not placed within 1 1/2 hours and shows signs of stiffening, remix it for about two minutes. This may restore workability. Discard the concrete if after remixing it is still too stiff to be workable. Do not add water to concrete that has stiffened beyond the point of workability.

HAND MIXING

Hand mixing should be done on a clean, hard surface or in a mortar box to prevent contamination by mud and dirt. A concrete slab makes a good working surface. The sand is spread out evenly on the slab, then the cement is dumped and evenly distributed. Mix the cement and sand thoroughly by turning them with a short-handled, square-end shovel until you have a uniform color, free from streaks of brown and grey. Streaks indicate that the sand and cement have not been thoroughly mixed.

Next, spread this mixture out evenly and dump the coarse aggregate in a layer on top. The materials are again turned by shovel until the coarse aggregate has been uniformly blended with the sand and cement. After at least three turnings, form a hollow in the center of the pile and slowly add the proper amount of water. Finally, turn all the materials in toward the center and continue mixing until the water, cement, sand, and coarse aggregate have all been thoroughly combined.

PREPACKAGED MIXES

Most building materials suppliers, hardware stores, and even some supermarkets sell prepackaged concrete mixes. All the necessary ingredients—portland cement, sand, and coarse aggregate—are combined in the correct proportions. The most common packages are 45- and 90-pound. A 90-pound package makes 2/3 cubic foot of concrete. A 45-pound package makes 1/3 cubic foot. All you do is add water and mix. Directions for mixing and adding water are given on the bag.

PLACING THE CONCRETE

When the concrete has been thoroughly mixed, or as soon as the readymix arrives, everything should be in readiness for placing. Although each job is different, most applications will employ the following tools: square-end shovel, straightedge or strike board, bull float or darby, hand float, edger, groover, trowel, broom, and garden hose. You will also need a wheelbarrow or chute to get the mix where you want it to go. Some sort of curing materials should also be at hand (see Chapter 1).

Make sure that you have a method of getting the mix to the forms. If the readymix truck can be backed right up to the site, your problem is solved. The chute on the truck will extend the reach by up to 10 feet. Otherwise, you will need a wheelbarrow and a solid pathway to wheel it on. If the ground is mucky or rugged, make sure you place boards down.

Never try to wheel concrete up a steep grade. A small grade here and there can be managed, particularly if you can get some sort of start. But trying to push a wheelbarrow uphill from a standing start is a bigger job than most of us can handle. If you suspect there will be difficulty in getting the readymix to the site, explain the problem to the dealer. He may be able to suggest a solution. If not, your only choice is machine mixing close to the site.

Using the chute or wheelbarrow, place the concrete in the forms to full depth, spading along the sides to complete the filling. Try to load the concrete as close as possible to its final position without too much dragging and shoveling. Start in one corner and continue dumping until you reach the other side. Use the shovel and the concrete hoe to get as uniform coverage as possible (Fig. 4-2). Push, don't pull, with the shovel, to prevent separation.

When you have poured enough concrete to fill the forms and compacted it along the sides of the forms, the next few operations—*striking off* and *rough floating*—should follow immediately. A prime requisite for successful finishing is that the rough floating be completed before *bleed water* starts to form on the surface of the concrete.

To insure that finishing is completed before the excess water comes to the surface, only pour as large a section as can be worked at one time. Pour the concrete in stages if the job is too large for the workers on hand.

STRIKING OFF AND FLOATING

After placing the concrete, strike off the surface with a *darby* or 2 × 4 straightedge, working it in a sawlike motion

Fig. 4-2. The concrete is moved to fill the forms with a concrete hoe or square-ended shovel as shown.

Fig. 4-3. Concrete is struck off with a piece of 2×4 or with a strikeoff board specifically made for the occasion.

Fig. 4-4. For a fast, professional strike off job, a vibrating strike off machine can be rented in some areas.

Fig. 4-5. The bull float is tilted slightly away from you as you push it forward, then held flat as you pull it back.

across the top of the form boards. If you'll be doing a lot of concrete work, make your straightedge as shown in Fig. 4-3. For an even better job, a vibrating *strike off* can be rented that also compacts the concrete (Fig. 4-4). The strike off or *screeding* action smooths the surface while cutting off the concrete to the proper elevation. Go over the concrete twice in this manner to take out any bumps or to fill in low spots. Tilt the straightedge slightly in the direction of travel to obtain a better cutting effect.

Immediately after striking off, the surface is rough-floated to smooth it and remove irregularities. Use the small wood hand float for small or close work, and the large bull float for larger areas. The darby is an excellent all-around tool that can be used as a straightedge as well as a float. The bull float is tilted slightly away from you as you push it forward then flattened as it is pulled back (Fig. 4-5). The darby is held flat against the surface of the slab and worked from side to side.

Do not overdo any of the preceding motions. Overworking of concrete causes excessive water and fine particles to rise to the surface, which renders the surface more prone to flaking and chipping. If holes or depressions are left after floating, add more concrete and float over again.

63

5

The Finishing Process

If you use air-entrained concrete, the finishing process can begin almost immediately after rough floating. You do not have to wait long on a hot, dry, windy day. If the weather is cool and humid, however, you may be forced to wait several hours. The key to proper timing is whether or not there is *water sheen* on the surface. Begin when the sheen has disappeared, which is quicker on days when more evaporation can be expected.

Another test is to step on the concrete briefly. The identation from your shoe should be no more than 1/4 inch deep. The mark will be removed during final floating.

Ordinarily, the surface should be ready for finishing by the time you have finished cutting the concrete away from the forms. This is accomplished by working a pointed trowel along the inside of the forms to a depth of about one inch (Fig. 5-1).

The first finishing step is *edging* which should take place as soon as the surface is stiff enough to hold the shape of the edging tool. Edging produces a neat, rounded corner to prevent chipping and damage, which could be a problem once the forms are removed. This operation also compacts and hardens the slab surface where floating and troweling is least effective.

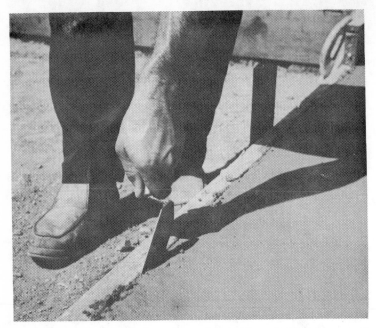

Fig. 5-1. As soon as the concrete is poured and rough-floated, you may begin cutting it away from the forms with a pointed trowel.

For slabs, a stainless steel edger with a 1/2-inch radius is recommended. The edger is run between the forms and the concrete, with the body of the tool held almost flat on the concrete surface (Fig. 5-2). When moving the edger forward, the leading edge should be tilted slightly upward. When moving in the opposite direction, the other end is tilted up. Be careful not to let the edger sink too deeply into the concrete, since deep indentations may be difficult to remove with subsequent finishing procedures. It may be, especially for a beginner, that the process will have to be repeated again once the other steps are concluded.

CONTROL JOINTS

Control joints are desirable whenever a slab covers more than 10 feet in any direction. These joints help keep cracks from starting, and they keep cracking from extending all the way across the slab if it does occur. Jointing, or grooving, is done after edging, or simultaneously if you have enough helpers.

Like other construction materials, concrete contracts and expands slightly under varying conditions of moisture and temperature. When new, and when hot and wet, concrete attains its largest volume. When dry and cold, it contracts. These changes are normal in all concrete, but unless provisions are made to control them, cracks may result. Other joints used to control cracking are *isolation joints*, installed before placing, as described in Chapter 3, *control joints*, and *construction joints*.

Control joints, sometimes called *contraction joints*, are made with a hand tool, sawed, or formed by using wood divider strips. The tooled or sawed joints should extend into the slab 1/4 to 1/5 of slab thickness. A cut of this depth provides a weakened section that induces cracking to occur beneath the joint where it is not visible.

In driveways and sidewalks, control joints should be spaced at intervals about equal to the slab width. Drives and walks wider than about 10 to 12 feet should have a longitudinal control joint down the center.

Fig. 5-2. Edging begins when the concrete is stiff enough to hold the shape of the edging tool.

Spacing of control joints in patios should not exceed 10 feet in either direction. The panels formed by control joints in walks, drives, and patios should be as square as possible. Long, thin panels crack more easily. As a general rule, the smaller the panel, the less likelihood of random cracking. All control joints should be continuous, not staggered or offset.

Hand tools for jointing are called *groovers* or *jointers*. Like edgers, they are made of stainless steel and other metals and are available in various sizes and styles. The radius of a groover should be 1/4 to 1/2 inch. The *bit* or cutting edge should be deep enough to cut the slab to a minimum of one-fifth, and preferably one-fourth, of the slab depth. Groovers with wornout or shallow bits should not be used, but saved for making decorative finishes. (These are detailed in the next chapter).

Mark the location of each joint with a string or chalk line on both side forms and on the concrete surface. A straight piece of 1 × 6 or larger should be used to guide the groover. The board should rest on the side forms and be at right angles to the edges of the slab. The groover is held against the side of the board as it is moved across the slab (Fig. 5-3).

Push the groover into the concrete and move it forward while applying pressure to the back of the tool. After the joint is cut, the tool is turned around and pulled back over the groove to provide a smooth finish. If the concrete has stiffened to the point where the groover will not penetrate easily to the proper depth, a hatchet can be used to push through the concrete. The groover is then used to finish the joint. All joints should be 1/4 to 1/5 the thickness of the slab, or about one inch deep.

Control joints can also be cut with a concrete saw. For small jobs, a power circular saw equipped with a masonry cutting blade may be used. A sawed joint must be cut as deep as a hand-tooled joint. Sawing should be done as soon as the surface is firm enough not to be torn or damaged by the blade, four to twelve hours after the concrete hardens.

A *construction joint* is necessary whenever a slab is poured in sections. Avoid this if possible; but, if sections are necessary, plan and locate them so that their intersections act

Fig. 5-3. Control joints are made with the groover, using a straightedge for a guide.

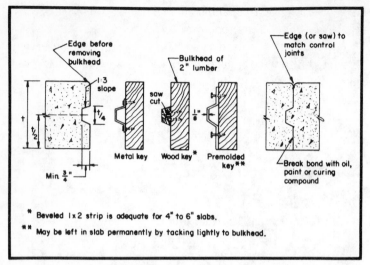

Fig. 5-4. Details for making a construction joint.

as control joints. A *keyway* is required to provide load transfer across a construction joint. This insures that the sections will remain level. The keyway is formed by fastening metal, wood, or premolded key material to a wood bulkhead. Concrete above the joint should be hand tooled or sawed to match the control joints in appearance (see Fig. 5-4).

FINAL FLOATING

Following edging and jointing, the surface is again floated. This embeds large aggregate just below the surface, removes imperfections left in the surface by previous operations, compacts the concrete, and consolidates mortar at the surface for any further finishing operations.

Hand floats are used for this operation. Magnesium floats are light, and strong, and they slide easily over a concrete surface. They are recommended for most work, especially for air-entrained concrete. Wood floats drag more, and hence require greater effort; but the wood produces a rougher texture. This may be preferred when good skid-resistance is required and floating is used as the final finish.

The hand float should be held flat on the concrete surface and moved with a slight sawing motion in a sweeping arc to fill in holes, cut off lumps, and smooth ridges. Professionals often use power floats or *compactors*.

When floating is used as a final finish, it may be necessary to float the surface a second time after some hardening has taken place. This will impart the desired final texture to the concrete (Fig. 5-5).

Because marks left by edgers and groovers are removed during this floating, the edger or groover will have to be rerun after final floating if its marks are desired for decorative purposes.

TROWELING

Immediately after final floating, the surface may be *troweled*. This produces a smooth, hard, dense surface. Troweling should never be done on a surface that has not been floated a second time; bull-floating or darbying is not sufficient preparation for this step.

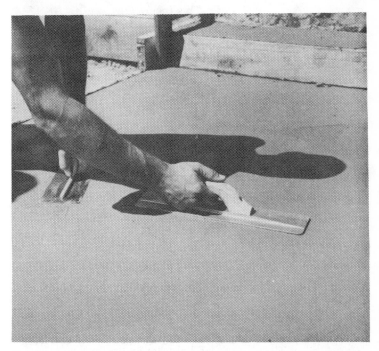

Fig. 5-5. A magnesium hand float works best for final floating, and produces a nice, smooth surface which usually does not need troweling.

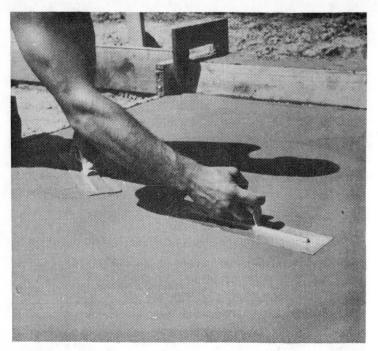

Fig. 5-6. Troweling produces a smooth, hard, dense surface. The trowel is held flat on the surface and moved in a sweeping arc on the first pass. If more troweling is necessary, the blade is tilted slightly.

On large slabs where you cannot reach the entire surface, you must work on kneeboards. When hand finishing, you can float and immediately trowel an area before moving the kneeboards. This operation should be delayed until the concrete has hardened enough that water and fine material are not brought to the surface. Too long a delay, however, will result in a surface that is too hard to finish. Premature finishing may cause scaling or dusting.

Hand trowels are made of high-quality steel in various sizes. Normally, at least two sizes are used. For the first troweling, one of the larger tools—18 by 4 3/4 inches, for example—is recommended. Shorter and narrower trowels are used for additional trowelings as the concrete sets and becomes harder. For the final troweling, a 12 × 3-inch trowel, known as a *fanning trowel*, is recommended. When one trowel is used for the entire operation, it should measure about 14 × 4 inches.

For the first troweling, hold the trowel blade flat on the surface (Fig. 5-6). If it is tilted, it will form ripples that are difficult to remove later without tearing the surface. Use the hand trowel in a sweeping arc motion, each pass overlapping one-half of the previous pass. In this manner, each troweling covers the surface twice. The first troweling may be sufficient to produce a surface free of defects, but additional trowelings may be used to increase smoothness and hardness.

Allow some time after the first troweling to permit the concrete to become harder. When only a slight indentation is made by pressing your hand against the surface, the second troweling should begin; use a smaller trowel held with the blade tilted slightly. The final pass should make a ringing sound as the tilted blade moves over the hardening surface.

If you have a large slab, you may want to consider renting a professional power trowel. If you use one, however, you will need at least one additional pass with a hand trowel. Troweling produces a very smooth surface, which can be slippery. Some masons automatically broom the surface after troweling to produce a non-skid surface, but this is not always necessary. Brooming is discussed as a decorative finish in Chapter 7.

6

Concrete Curing

The chemical reaction between cement and water is called *hydration*, a process that must continue for several days to a week after placing the concrete, in order to attain maximum durability. If too much water is lost by evaporation, the chemical reaction ceases. The same is true when temperatures get below 50° Fahrenheit. Hydration slows almost to a standstill as the temperature approaches the freezing mark.

Curing is a vital step in concrete work. It is essential for keeping water in the concrete for the right length of time. As soon as the finishing process is complete, and the surface is hard enough so that it will not be damaged, curing should begin. In warm weather, the curing process must continue for five days. For every 10 degrees less than 70°, add an extra day (six days at 60 degrees, seven at 50 degrees). At no time should the temperature of the concrete fall below 50 degrees. If the weather gets colder than that, other steps must be taken. These are described below.

STANDARD CURING METHODS

The recommended curing method for the homeowner is to keep the concrete surface damp by applying wet burlap. Rinse out the burlap before use, particularly if it is new, and spread

Fig. 6-1. The recommended curing method for homeowners is to cover the concrete surface with wet burlap.

it over the slab (Fig. 6-1). The burlap should be checked several times a day to see that it does not dry out. Periodic sprinkling, at least daily, is the best method of keeping the burlap continuously moist.

Another method of keeping the surface wet is to run a sprinkler or soaking hose continuously over the surface. Never let the surface get dry at any time, since partial curing will ruin the surface.

For small jobs, try *ponding*, or building sand or earth dikes around the edges of the slab. The dikes keep a pool of water on the surface. The water must be deep enough to cover the entire surface of the concrete and to prevent dry spots.

A popular method used by concrete contractors is to cover the entire surface with a moisture barrier such as plastic sheeting or waterproof paper (Fig. 6-2). Further moisture additions are not necessary. This is preferred by professionals because they don't have to come back and soak the area periodically—a time-consuming and costly nuisance. This is not the case with the homeowner, who is around often enough to keep the wet-curing process under easy surveillance.

The problem with plastic sheeting is that it must be laid perfectly flat. The material must be thoroughly sealed at joints and anchored firmly on all sides. If the plastic or paper is allowed to wrinkle, patchy discoloration results. This might not be bothersome on some projects, but it is unsightly in a show area such as a patio.

An excellent, though expensive, curing method is to spray a curing compound. This is recommended when you want colored concrete, since the compound may be pigmented, thus accomplishing two jobs at once (see the next chapter for other coloring methods). Complete coverage is essential for the compound to work properly, and a second coat, applied at

Fig. 6-2. Plastic sheeting is preferred by contractors, because further care is not necessary. But it's a tricky method.

right angles to the first, is advised. Curing compounds should not be used in late fall in the northern states or any other areas where deicers are used to melt snow and ice. The curing compound may prevent proper air-drying of the concrete, lowering its resistance to scaling caused by deicers.

HOT AND COLD WEATHER TIPS

Hot Weather

Procedures for curing must differ according to weather patterns. When the weather is very hot, evaporation of water from the concrete mix takes place rapidly. This may cause cracking and finishing problems. The concrete mixture should never be allowed to rise above 90° Fahrenheit. Start your projects in the late afternoon or very early morning to avoid the worst of it. It also helps to sprinkle the aggregate with water and to chill the mixing water with ice (making sure it is melted before placing the concrete).

Other preventive measures include:

- dampening of subgrade and forms
- keeping down finishing time by having extra manpower
- erecting sunshades and windbreaks of canvas or polyethylene
- covering the pour temporarily with curing materials during finishing, and uncovering only that portion being worked on
- spraying lightly while working
- starting curing immediately after finishing.

Cold Weather

Cold weather concreting should be avoided if there is any way to do it, but if you can't, there are ways to proceed. Adding three per cent calcium chloride to the mix is one way, but this is not highly recommended by the Portland Cement Association because of problems that have occurred. If you do use calcium chloride, it should be added to the mix in the form of a solution. You may have heard that antifreeze compounds or rock salt will do the trick, but they won't. These materials will reduce the strength and wearing qualities of the concrete.

The best way of minimizing cold-weather problems is to order preheated readymix. It is virtually impossible to heat your own mix, so stick with readymix in cold weather. It is a good idea, also, to use high-early-strength concrete, which is made by mixing Type III (High-Early-Strength) cement, or by adding an extra 100 pounds of cement for each cubic yard. This type of concrete should cure in three days instead of seven.

The concrete must be kept warm after finishing, too. The way to accomplish this is to cover the slab with several layers of insulation. There are several ways of doing this. The best method consists of placing a layer of building paper, followed by a foot or two of straw, then covering everything with plastic sheeting to keep it dry and in place. Slab edges and corners are the most vulnerable areas and should be fully protected. If the cold is severe, some sort of artificial heat should be applied. The insulation should remain in place for about a week.

7

Special Finishes

Most people are quite satisfied with the look of plain concrete. For the usual basement floor, foundation, driveway, sidewalk, or similar utilitarian purpose, the concrete mix described in the previous chapters will be perfectly satisfactory. Many do-it-yourselfers, in fact, are not aware of the myriad possibilities of decorative concrete. If they want a colored patio, for example, they slap on a coat of cheap concrete paint.

Concrete paint is probably the worst way to achieve a different effect. Cheap paint wears off and peels in a short time, and even good paint doesn't last forever. The result, in either case, is a concrete surface that looks very unsightly. One cure is to remove the paint that is left and start over, but this is an onerous and self-defeating task.

If the look of the standard concrete mix is not pleasing to you, there are a lot of options. You can, for example, use a different building material. But you don't *have* to. There are many things you can do to concrete to spruce up its appearance. Here are some of them:

- color it (not with paint)
- texture it
- expose the aggregate (Fig. 7-1)
- give it a slush coat (for walls only)
- cover it with another material (for walls only)

Fig. 7-1. Concrete needn't be dull gray. Here, exposed aggregate makes an interesting and inviting drive and entrance.

COLORING CONCRETE

One of the easiest methods of coloring concrete is to use a pigmented curing compound, as was described in Chapter 6. This method combines two operations in one. There are other methods, too.

Using Mineral Pigments

If you want lasting color in concrete, this is the most effective procedure. The pigments are relatively expensive, but in the long run, infinitely preferable to paint. Not every color is available, but there is enough variety to satisfy most people. See Table 7-1.

Coloring can be used in two ways. The pigment may be added to the usual one-course mix, or the slab may be poured in two layers or courses, with the pigment only in the top course. If the slab is thin enough (four inches or less), the one-course method is probably better, even if you use more pigment. If the slab is thick, the two-course technique will save some money, but it means more work.

If you prefer a pure white color to the typical concrete gray, your problem is easily solved. Just use white portland cement. This costs a little more, but very little by comparison, and no special mixing techniques are necessary. The one-slab method is obviously best here.

You should use white portland cement for most colors anyway. Except for black or dark gray, white cement will yield better and brighter colors. Regular cement is okay for black and dark gray.

In the one-course method, a specified amount (ask your dealer) of mineral oxide is added to the concrete mix. All mix materials must be controlled by weight to assure uniform color, and the pigment should be thoroughly mixed before the water is added. A separate mixer from regular concrete should be used to avoid streaking, and all tools must be thoroughly cleaned before working on the colored mix.

When using the one-course technique, the subgrade should be thoroughly soaked to prevent coloring agents from leeching into the ground. Soak the subgrade well the night before placing the concrete, or use a moisture barrier over the subgrade. Finish as with regular concrete.

The two-course method is similar to the first, except that the first course is applied the same as any other slab. The surface is left deliberately rough to provide a good bond for the second course. If it looks too smooth, *scarify* or scar the unhardened surface to insure a good bond. This is done by scraping the surface with a sharp tool. The first course should be about 1/2 to one inch below the final grade.

Table 7-1. Guide to Coloring Compounds

Desired Color	Material to Use
white	white portland cement, white sand
blue	cobalt oxide
brown	brown oxide of iron or burnt umber
buff	yellow oxide of iron or yellow ocher
green	chromium oxide
red	red oxide of iron
pink	red oxide of iron (smaller amount)
dark gray	black oxide of iron

Fig. 7-2. Pigmented dry shake may be applied before final finishing, according to manufacturer's instructions.

The second course is colored in the same way that concrete for one-course coloring was treated, and laid down after the base coat stiffens and clears of surface water. This course is then finished the same way as any other concrete surface.

Using Dry Shake

Dry shake coloring is applied after the concrete is floated, edged, and grooved. Follow manufacturer's instructions, first applying about two-thirds of the total dry shake, then floating, edging, and grooving again (Fig. 7-2). The remainder is then applied, and the surface is finished again. Make sure that the dry shake is worked thoroughly into the concrete surface to a uniform color. Troweling follows in the usual manner, if desired.

Using Concrete Stain

Used like paint, concrete stain actually does stain the concrete surface, instead of merely covering it as paint does. Solvent-type stains are similar to wood stains and, as a matter of fact, you *can* use wood stains on concrete. Those made specifically for concrete are better, however, and can be applied to old concrete as well as new.

For new concrete, organic stains such as Kemiko are best. They come in brown, black, rust, beige, and green, and react chemically with the concrete, causing a long-lasting colored finish. Follow manufacturer's directions for the type of stain purchased.

Using Paint

If, in spite of warnings, you must use paint, at least buy the right type. Exterior latex is all right for walls and floors with little traffic. Portland cement paint is recommended for indoor walls. Floors subjected to heavy traffic should be covered with an epoxy paint. Chlorinated rubber-base paints are best for all-around applications.

TEXTURING CONCRETE

The cheapest and easiest way to add interest to your concrete is to give it a special texture. There are many ways to do this, with the most common being a *broomed* finish. Brooming follows the last finishing operation, and can be accomplished by using whatever type of broom suits your fancy. Professionals use a broom specifically made for concrete texturing purposes (Fig. 7-3), but the one-time user can do just as well with a regular household broom.

For a soft broomed finish, use a soft-bristled broom on a troweled concrete surface. Coarser textures are produced with a stiff-bristled broom on newly floated concrete. A variety of patterns can be used—straight, curved, wavy, circular, or swirled (Fig. 7-4). Brooms should be dampened each time they are passed over the concrete. The lines should run at right angles to the direction of traffic.

Swirling

Although a broom can be used for swirling, a more subtle effect is produced by running the float in a fan-like, semi-circular direction over finished concrete (Fig. 7-5). Pressure is applied as you swing your arm in an arc over the surface, overlapping the previous swirls. A coarser surface is supplied with a wood float, medium texture is supplied with a magnesium or aluminum float, and the finest texture is supplied with a steel trowel.

Fig. 7-3. Brooming provides an interesting and nonslip surface.

It may take some time for you to get the desired effect without botching up the job. Use your arm instead of your wrist to achieve a uniform arc. Let the surface dry a little more than usual before curing so that you don't mar the texture.

Travertine

Keystone or pitted finishes are recommended only where hard freezes do not occur. The *travertine* finish creates holes

Fig. 7-4. A wavy, broomed surface takes a little more skill than a straight one, but can be mastered with some practice.

Fig. 7-5. The swirled surface is done with a float, swinging the arm in an arc and overlapping the previous swirls.

Fig. 7-6. The travertine finish shown was made by embedding rock salt in the top of the concrete surface.

or depressions in the surface which resemble travertine marble. The problem in freezing areas is that the depressions collect water, which can spall the surface if it freezes.

The more elegant travertine finish is obtained by mixing a slush coat of mortar, usually pigmented light yellow. The mortar is applied in a splotchy manner with a brush so that ridges and depressions are formed. When the mortar has hardened slightly, it is troweled down to flatten the ridges, giving a smooth finish in the high areas and a mottled effect to the depressed spaces.

Rock salt imparts a similar surface, but the surface is pitted rather than mottled (Fig. 7-6). The salt is pressed or rolled onto the surface so that only the tops of the grains are exposed. After hardening, the surface is washed and brushed, dislodging the salt and leaving pits or holes.

Patterns

Only your imagination limits the type and number of geometric patterns that are possible to create on concrete. You can make the concrete resemble flagstone (Fig. 7-7), brick, or a Mondrian abstract painting. The best tool for this is a piece of Z-shaped copper tubing, either 1/2 or 3/4 inch in diameter. The surface is scored immediately after rough floating, since it must be plastic enough for the tool to push the aggregate aside. A second pass is made after finish floating to

Fig. 7-7. Geometric patterns of any kind may be created by using a piece of copper tubing to score the still-plastic surface.

Fig. 7-8. Special stamping tools are used to create brick and other novelty effects.

smooth out the joints. A stiff bristle brush will touch up any areas which have burrs or ridges.

Another method is to make one-inch strips of 15-pound roofing felt and install them in the surface after rough floating. The strips are laid on the top in the desired pattern, patted flush, then floated over. Color can be added with dry shake afterwards, and the grooves can be left natural to resemble mortar. The strips are removed after curing. Grooves can actually be filled in later with mortar if you want a really authentic look.

Patterns can also be stamped into the surface with special stamping tools, as illustrated by the brick design shown in Figs. 7-8 and 7-9. The stamping tools shown and other patterns may be available on a rental basis from your concrete dealer or rental shop. You should do the stamping following the final finishing, but don't trowel the surface more than once. At least two large pads are necessary, and more are better for big surfaces. One pad is placed next to the other to insure correct alignment, then the operator steps from one onto the other to impress the pattern into the surface to a depth of about one inch (Fig. 7-8). If foot pressure is insufficient, a hand tamper is used. A small hand pad, tapped with a hammer, is used to complete patterns near the edges. For a brick pattern, a regular brick jointer is used to create authentic-looking joints (Fig. 7-9). When using stamping tools, a mix of smaller aggregates is used, the largest being the size of pea gravel. Watch your timing if using this procedure; stamping should be completed no more than five hours after placing so that the surface doesn't get too hard. If working on a large slab, get plenty of extra tools and extra hands to work them.

DIVIDER STRIPS

As discussed in Chapter 3, wood divider strips serve both as decoration and control joints. These can be placed in any number of patterns, and can be made of materials such as metal, brick, concrete masonry, stone, or plastic. Although masonry strips can be set in a sand bed as described in subsequent chapters, they are best set in mortar for use with concrete.

One advantage of divider strips is that they segment the concrete in certain areas, which means that each section can

Fig. 7-9. Use a brickmason's jointer to dress up indentations and create some artificial defects for better realism.

be poured and finished separately without having to use construction joints. If you wish to work at a more leisurely pace, or must contend with a lack of helpers, divider strips can serve a definite utilitarian purpose, as well as being highly decorative.

EXPOSED AGGREGATE

One of the most satisfactory of all concrete finishes is exposed aggregate, sometimes called *terrazzo* (usually incorrectly). Terrazzo is actually a separate technique which utilizes half-inch thick topping over a base slab. True terrazzo contains decorative and more expensive aggregates such as marble, quartz, or granite chips. It is a difficult job for the handyman and is usually installed by a qualified terrazzo contractor.

Exposed aggregate work can be performed by the handyman, but it is not easy. Any "exposed-ag" surface is highly durable and anti-skid, as well as very attractive, however.

You can get an exposed-ag surface by ordering a conventional mix, but with a high proportion of coarse-to-fine aggregates and a low slump (one to three inches), so that the coarse aggregate stays near the surface. The usual procedures are followed, except that floating is done gently to keep the large stones from being pressed down too far.

When the water sheen disappears, and the concrete can stand a man's weight without indentation, the aggregate is ready for exposing. The surface must be washed and broomed simultaneously. If the stones become dislodged or overexposed in the process, wait a little while longer and try again.

A much more satisfactory method for the do-it-yourselfer is called *seeding*. The concrete mix is placed in the usual manner, except that it should be leveled off at 3/8 to 1/2 inch below the top of the forms to allow for the extra aggregate. Strike off and float in the usual way, then spread rounded stones evenly over the surface with a shovel (Fig. 7-10). Fill in the bare spots by hand until the surface is completely covered with aggregate. If the first few stones sink to the bottom, wait a half-hour or so until the mix gets a little stiffer.

When you have a good, even stone cover, tap it into the surface of the concrete with a 2 × 4, darby, or wood float. Then go over the entire surface with your wood hand float, working the stones down into the concrete until they are entirely covered again by cement paste (Fig. 7-11). The surface will look almost like it did before.

Wait for an hour or two, until the slab can bear the weight of a man on kneeboards without leaving an indentation. Then brush the surface lightly with a stiff nylon-bristle broom to remove the excess mortar.

The final, and most difficult, job is hard-brooming the surface while washing the stones with a fine spray. Either have a helper for this, or alternately spray and broom. But

Fig.7-10. Spread rounded aggregate evenly over the surface until it is completely covered with stone.

Fig. 7-11. After tapping the stones into the concrete with a darby or 2×4, go over with a wood float, working the stones down until they are entirely covered by concrete.

brush hard enough to dislodge as much cement film as you can. You may be able to rent a special exposed-aggregate broom with water jets as shown in Fig. 7-12. Ideally, all you should see are the pretty, colorful stones. If the surface is too dull, give it a bath with muriatic acid.

NOVELTY FINISHES

The only limitation when finishing concrete is an infertile imagination. Almost any pattern is acceptable as long as you can figure out a way of applying it.

Reasonably *non-skid* surfaces are provided by using the hand float without troweling, as described in the previous chapters. Even better is brooming or using exposed aggregate. For areas around pools or other places where there is a real danger from wet and slippery surfaces, it may be advisable to use long-lasting abrasive grains such as silicon carbide and aluminum oxide.

If you want a really classy surface around your pool for nighttime frolicking, silicon carbide grains will give you "sparkling" concrete. The sparkling effect is particularly effective under the artificial light. Aluminum oxide is just as effective as silicon carbide for imparting an abrasive surface, but it does not sparkle as silcon carbide does. Silicone carbide is black, while aluminum oxide comes in gray, brown, or white.

Abrasive grains are applied very much like dry shake. The grains are spread uniformly over the surface, about 1/4 to 1/2 pound per square foot, and lightly troweled into the surface (Fig. 7-13). Check manufacturer's instructions.

Concrete is a highly durable product, and once it is properly mixed and placed, you can do anything you want to the surface, within reason. One tricky, but interesting flourish,

Fig. 7-12. The final step is brushing the stones while washing with water. A rented broom made to combine these operations will facilitate the job considerably.

Fig. 7-13. Silicon carbide grains embedded in the surface give the surface a dark, sparkling effect while also providing an abrasive, non-skid surface.

is to embed leaves into the surface. The leaves should be embedded firmly, but there should be no cement paste on top. After the concrete has set sufficiently, the leaves are removed. Thorough curing is essential for this type of delicate figuration.

Another possibility is to make circles with variously sized tin cans, squares, and rectangles with stiff boxes or other implements. The artist may draw his own designs with a stiff brush or a pointed tool. There is no technical limit, but there may be an aesthetic one. A truly far-out or macabre design may turn off visitors and prospective buyers. Remember that your design will be there for a long time.

8

Walls,
Foundations, and Steps

Slabs are usually simple for the handyman to construct. Any project requiring more complicated and sturdy forming is not recommended as a *first* project. These projects can be done, of course, by following instructions carefully, but you really should try your luck with horizontal applications before attempting the work detailed in this chapter.

Concrete is heavy stuff. When placed on a slab, the pressures on the forms are relatively light. As the project gets vertically higher, however, the pressure on the forms becomes increasingly greater. Hence, the need arises for exact formwork with extensive bracing. If the concrete is to support other heavy work, as in a home foundation, the need for a good sturdy job increases dramatically. If your wall is to hold back earth and water, as with a retaining wall, strength of construction is again extremely important.

Steps present other problems. The forming is rather complex, and finishing can be difficult. Again, the forms contain a lot of concrete, and must be built to withstand enormous pressure.

FOOTINGS

Any vertical concrete job exerts a lot of downward pressure even after it has set and cured. Walls and steps which

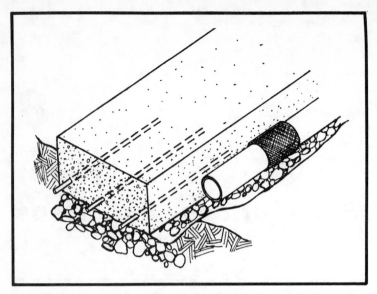

Fig. 8-1. A simple footing for low walls. Reinforcing bars may or may not be necessary.

are poured without a *footing* can and often do sink into the ground. Footings are designed to diffuse the downward pressure and provide a solid base for the structure above (Fig. 8-1).

Before building any type of wall, building codes should be carefully checked. There will probably be specific requirements for footings and reinforcement. These specifications must be met in your project. If there are no code specs, use the following guidelines:

- Footings should be at least two feet deep, or six inches below the frostline.
- Footings should be at least twice as wide as the wall itself, or six to eight inches wider on all sides for steps.
- The depth of the footing should be the same as the width of the wall.
- Footings should rest on a six-inch bed of crushed rock.
- Reinforcing bars may be required in many cases. If not specified by the code, ask your concrete dealer about your particular application. Footings usually require three horizontal bars inside the footing itself

(about one-third from the bottom) or vertical bars set on 24-inch centers into the footing and extending from it into the wall itself. (See next section for use of reinforcing bars with block walls.)

- Most footings require a *keyway* in the top surface. This insures a good bond between the footing and the wall and helps prevent sideways movement. A keyway is formed by inserting a tapered oiled 2 × 4 into the top of the footing surface. Remove after setting.

- All walls must have drainage provisions. If there is no drainage in the wall itself, install drain tile along the footings.

Footings need not be pretty, but they must be straight and level. Use batter boards and lines to lay out the design, and dig down to the required depth. Grade carefully and lay in your stone base. If the earth is firm, and you have dug reasonably straight sides, no forming may be necessary. If using the earth for forming, make sure that the sides are level.

When the earth is soft or sandy, forms will be needed. Most footings will require only one or two boards, but if you form much higher than six or eight inches, bracing will probably be required, and formwork will approximate that of regular walls as described below. Horizontal reinforcing rods can be placed on flat stones. Vertical rods are tied to horizontal ones and bent with the ends extending into the side forms (Fig. 8-2). Drill holes in forms to accommodate these rods. If using vertical rods, the keyway must be cut in half and installed on each side of the rods.

It is doubtful that the footings will be more than two feet in depth, but if they are, make two pours, tamping down the first pour before pouring the balance. Make sure that the first section does not harden before you pour the second.

Rough-floating is all that is required for footings, but the footing should be cured in the same manner as a slab.

WALLS

Again, check building codes. Vertical reinforcing rods are often required, and sometimes horizontal rods or hardware cloth (thick wire mesh) are needed as well. Horizontal rods

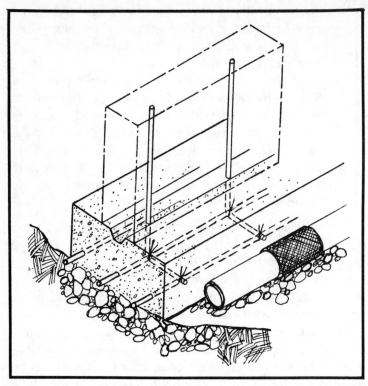

Fig. 8-2. A keyed footing with vertical reinforcing bars for high walls or other walls that undergo high stress.

are tied to vertical rods with steel wire. Rods come in lengths of up to 20 feet. If longer ones are necessary, they should be overlapped 18 inches. Tie all rods together where they cross. Rods should be #3 or #4 (3/8- to 1/2-inch thick). Hardware cloth comes in five- or six-foot widths. It should be placed at least two inches inside forms, and overlapped six inches where necessary. Cut with wire cutters and tie together with steel wire. See Fig. 8-3.

Forms are made of tongue-and-groove boards, plywood, or plyscord (plywood which is smooth on only one side). Bracing is done with 2 × 4s or even 4 × 4s if you are building a large retaining wall. Since walls rest on footings, it will be impossible in most cases to drive stakes into the ground next to the wall. Stakes are set a few feet outside the forms, and braces are formed in a triangular manner as shown in Fig. 8-4.

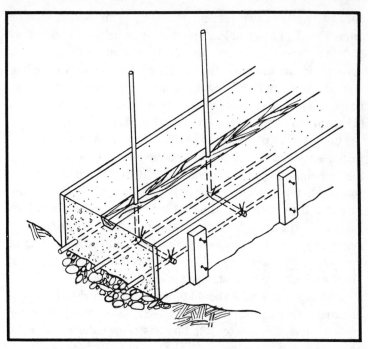

Fig. 8-3. How to make a keyed footing with vertical re-bars.

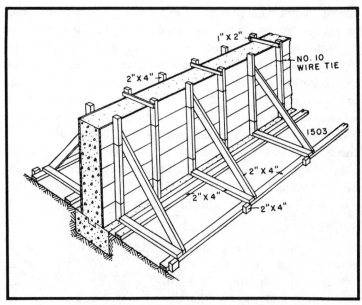

Fig. 8-4. Typical forming for wall on level ground.

Set braces about two feet apart, closer if the wall is high or thick. Cross ties are set intermittently to help hold the forms together. Wood spreaders should be placed, but not nailed, between forms to separate the forms. Tie wire or strong line around them, and pull them out after the concrete has been poured. Wires may also be strung between the two form sides to keep them spreading apart. These remain in place.

Concrete is poured into walls beginning at the ends and working toward the center. In no case should more than two feet of height be poured at one time. After that much has been poured, spade and tamp, then go on to the next batch, making sure that the concrete in the first layer does not harden before going on to the next. It is important here that there are enough helpers to complete the pour at one time, although this part of the job goes fairly quickly compared to pouring slabs. Formwork will be the toughest part of it.

The top of the wall should be struck off, checked for level, and rough-floated, but no further finishing is necessary. After the mix has started to set, any fastening devices such as anchor bolts may be installed. If wooden sills are to be placed on top of the concrete, it is probably easier to attach them with helical concrete nails. New concrete is fairly easy to nail into, but the job gets much more difficult as the concrete cures.

Retaining walls should not be constructed more than four feet high. Steeper grades should be stepped or terraced. One reason for this limitation is that higher walls are much harder to construct, but an even better reason is that for home applications high retaining walls are unsightly. If you must construct a wall higher than four feet, the bottom of the wall should be thicker than the top. This can be accomplished by either slanting the walls so that the bottom is wider (see Fig. 8-5), or by building the wall in steps, with the stepped portion into the hill and the straight part in front.

STEPS

Steps should be anchored to the foundation wall. If possible, anchors should be installed in the foundation in anticipation of the steps. If not, drill or chisel out a hole in the foundation, and mortar the ties. There should also be an isolation joint between the foundation and the steps.

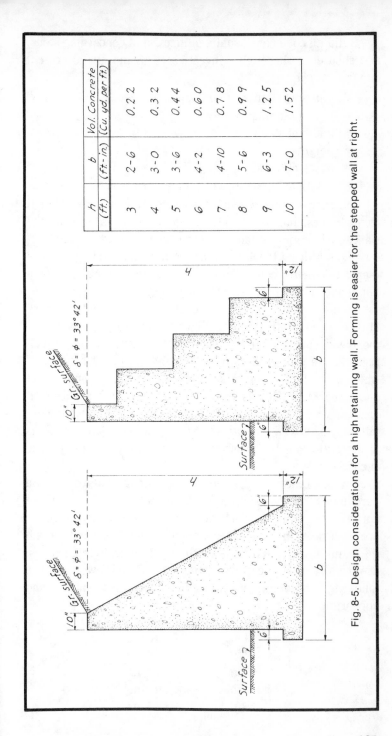

h (ft.)	b (ft.- in.)	Vol. Concrete (Cu. yd. per ft.)
3	2-6	0.22
4	3-0	0.32
5	3-6	0.44
6	4-2	0.60
7	4-10	0.78
8	5-6	0.99
9	6-3	1.25
10	7-0	1.52

Fig. 8-5. Design considerations for a high retaining wall. Forming is easier for the stepped wall at right.

Building codes are usually quite explicit concerning steps, and should be thoroughly investigated. Some of the critical dimensions are:

- height of flights without landings
- size of landings, if any
- width of steps
- height of risers
- depth of treads
- relationship between risers and tread size

In general, steps are 48 inches wide. They may be as small as 30 inches, but should always be at least as wide as the walkway leading to them and the door they lead to. Landings should be at least three feet long, with the top landing no more than 7 1/2 inches below the door threshold. There should be a landing for every flight five feet in height.

Smaller flights of less than 2 1/2 feet usually require 11-inch tread widths and 7 1/2-inch risers. For higher flights, the treads are usually at least 11 1/2 inches with six-inch risers. Note that both combinations add up to 17 1/2 inches, which is the figure regarded as the ideal combination of tread and riser size. This is the optimum relationship based on normal walking strides. The usual entrance stairs should follow this pattern for both safety and comfort. Safety reasons also dictate that the risers and treads should be uniform for each flight.

On the other hand, steps along a long walkway in the garden or other parts of the landscape should be built for eye appeal as well as comfort. Longer treads and shorter risers are preferable in these cases, although they, too, should retain the same proportion in any one flight. (Flights, in these cases, are usually only one or two steps at a time, however). The following combinations are often used for safety and appearance:

Riser	Tread
4 in.	19 in.
4 1/2 in.	18 in.
5 in.	17 in.
5 1/2 in.	16 in.
6 in.	15 in.

Stair footings may be made in the manner described previously, or two or more postholes may be sunk into the ground underneath the stairs to at least the same depth as the footings. The postholes should be six to eight inches in diameter and filled with concrete.

Novice step builders should try to confine steps to three levels if possible. Forming is similar to that for walls, except that cross-ties and wiring of forms together is neither practical nor necessary. The riser forms act as spreaders. Riser forms should be braced by running boards down and staking them in front, as shown in Fig. 8-6, or nailing boards between them, as shown in Fig. 8-7. Rigid bracing is a must. The higher the steps, the more rugged the bracing should be. Boards will suffice for smaller steps, as shown in Fig. 8-6, but for larger steps, you should really construct sidewalks made of 2×4 framing and plywood or hardboard panels as shown in Fig. 8-7. These walls, too, must be braced solidly.

A portion of the form construction depends on the finishing technique. In one method, the riser forms are stripped as soon

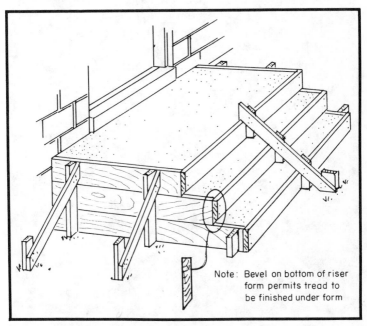

Note: Bevel on bottom of riser form permits tread to be finished under form

Fig. 8-6. Typical forming for entrance stairs to a house.

Fig. 8-7. Formwork for larger steps using panel side forms.

as the concrete sets enough so that the steps support their own weight. This can be 30 minutes to several hours depending on the weather and other factors. The surface is floated hard enough to bring mortar to the surface. Depressions are filled in with the excess or with new grout (see below). After the concrete sets a little more, it is troweled and brushed. Steps should *definitely* have an anti-skid surface.

A better method for most handymen, however, is to bevel the bottom edges of riser forms so that the treads can be completely finished with the forms in place (see Fig. 8-5). In this technique, the landing and the treads are finished in a method similar to that of slabwork, with the forms still in place. After the forms are stripped, defects are chipped off or filled with grout, as the case demands.

Wood cleats are used to attach riser forms to panel sideforms. If steps are built between walls and masonry walls, as is usually the case for basement stairs, there is nothing to nail to. Large wooden wedges will hold the risers in place in this event. Remember to allow for railing attachments, and to pitch risers about 1/4 inch per foot for drainage.

ORDERING CONCRETE

Concrete for walls should be similar to that used for slabs, except that a three-inch slump is better in this case. The same applies to steps, with the additional exception that the maximum-size aggregate particles should not exceed one inch in diameter.

Estimating the amount of concrete needed for steps can be difficult, particularly if fill is provided as discussed in Chapter 3. To compute overall dimensions, figure each step as a separate slab. If, for example, you have four steps, each 7 1/2 inches high, and four feet wide, consider that you have four complete slabs. If the landing is four feet long, and each tread 11 inches, you have the following:

top slab	$7\,1/2 \times 48 \times 48$ inches	=	17, 280	inches
second slab	$7\,1/2 \times 48 \times 59$ inches	=	21, 240	inches
third slab	$7\,1/2 \times 48 \times 70$	=	25, 200	inches
bottom slab	$7\,1/2 \times 48 \times 81$ inches	=	29, 160	inches
Total			92, 880	cubic inches

$$92,880 \div 1728 \text{ (cu. in./cu. ft.)} = 53.75 \text{ cubic feet}$$
$$53.75 \div 27 \text{ (cu. ft./cu. yd.)} = 1.99 \text{ cubic yards}$$

What complicates matters is the fill. The only real solution here is to estimate as closely as you can how much space you have in fill, then subtract it from the total cubic yards. Wall volume is estimated as if it were a vertical slab.

For large stairs, it is possible to use falsework inside to save concrete. False work is nothing more than inside forms that are left in place. Wood falsework should not come any closer than six inches from any face of the steps. It should be braced inside, or placed over fill so that the wood is supported by it. It is best to cover the false work with a waterproof membrane such as plastic sheeting to reduce moisture absorption by the falsework. This absorption can weaken the concrete, and cause swelling of the wood and subsequent cracking in the steps.

PLACING, FINISHING, AND CURING

Forms that will be removed quickly should be wet with water before placing the concrete. Any forms that will remain

Fig. 8-8. Raiser boards are removed and the risers are finished.

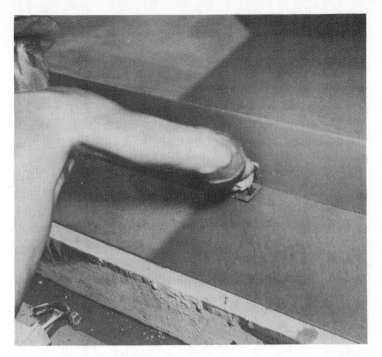

Fig. 8-9. Special step edgers are used for the corners.

for a few days should be oiled. Placing proceeds one step at a time. The bottom step is filled first and spaded against the forms. The next step up follows, and then the next. Each tread is struck off and floated as it is poured. The landing is struck off and floated last, as is any other slab.

Depending on which method is used—and sometimes there are compromises between the two—the forms are usually stripped away from top to bottom. With either method, it is possible to strip the risers as soon as the concrete is hard enough to hold a man on kneeboards (Fig. 8-8). Both risers and treads are re-floated, then edged along the forms. An inside edging tool is used where risers and treads meet (Fig. 8-9). If there is not enough cement paste to cover depressions, a little mortar consisting of one part cement to 1 1/2 parts fine sand may be applied. The entire surface is then troweled, and a damp brush run over the surfaces to get a fine, nonslip texture. Other nonskid methods, discussed in previous chapters, may also be used. This job must be done quickly enough so that the bottom steps do not set too firmly for proper finishing.

In any case, the side forms are usually left in place until at least the next day. If stripped early, the surfaces are floated and covered over with mortar to fill depressions. If left in place for a few days, the surface is saturated with water and mortar is then applied as above. The mortar is vigorously floated to fill all voids and the excess is wiped off. When dry, remove the dried mortar with a piece of wet burlap.

Steps are cured just like slabs, except that water-curing is the only effective method. Burlap is probably the easiest way to accomplish this, keeping it continuously wet as described in Chapter 6.

Wall forms should stay in place for at least several days, and up to a week, depending on weather conditions. They should be well oiled before pouring for easier removal. After the forms are removed, the wall may either be left as is, given a slush coat of mortar for better looks, or covered with some other material.

9

Casting—Your Own and The Dealer's

Precast concrete is made in forms and is also sold in many varieties. You can buy small brick-type castings for home and garden, precast steps or paving slabs (Fig. 9-1), or huge architectural castings for construction. Or you can make your own in any shape or form you wish.

Your concrete and masonry dealer should have precast concrete in many sizes and types—rectangular, round (Fig. 9-2), diamond-shaped, hexagonal—for use in patios and pathways. They also come in a variety of colors. Smaller units can be lifted and placed by hand, but sidewalk slabs, steps, and other larger units are usually set into place by a truck crane or by a vacuum lifting device which may be available from the dealer or manufacturer.

The most common precast units bought from a dealer are the smaller ones used for patios and walkways. These are best set in a bed of sand as discussed in a section on the brick-in-sand patio, in Chapter 16. Slabs used in streetside sidewalks and driveways are better set in a mortar bed, also discussed in Chapter 16.

Precast steps can be cantilevered from the foundation wall or set onto a footing. The footing can also be precast and come right with steps, or you can pour your own. Before you do

Fig. 9-1. Precast paving slabs make an attractive walkway.

Fig. 9-2. These precast rounds were made in the same way as regular exposed aggregate (see Chapter 7).

anything, consult the dealer to see what is available and what procedures must be followed. Most dealers will deliver the steps (they're very heavy) and set them in place with a crane (Fig. 9-3).

Fig. 9-3. Precast steps can be delivered to your door and installed on the spot.

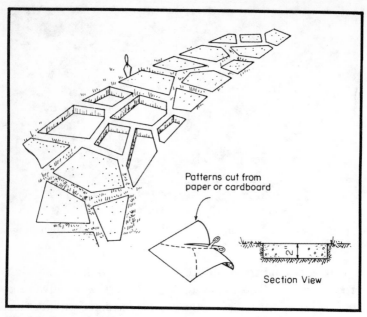

Fig. 9-4. Concrete "flagstones" are cast in place as two-inch slabs. Patterns make a more uniform-looking job.

IN-PLACE CASTINGS FOR THE GARDEN

Casting in place is similar to slab construction, except that the units are smaller, thinner (usually two inches), and not subject to a great deal of stress as are large slabs. Concrete "flagstones" are easy to make and considerably cheaper than the real thing (Fig. 9-4). If you want them to look more authentic, use one of the coloring methods discussed in Chapter 7.

There are two general methods of casting in place. One is to construct a re-usable form as shown in Fig. 9-5. Specific dimensions are given here, but you can use any other sizes that strike your fancy. All form boards (2 × 4s are best) should be beveled at the bottom for easier removal. Using screen-door handles also helps. The forms should be oiled before use and cleaned afterwards.

To use the type of form shown in Fig. 9-5, first excavate the area to a depth of two inches, then place the form inside the excavation. Concrete is then mixed in a proportion of one part portland cement, two parts sand, and 2 1/4 parts coarse

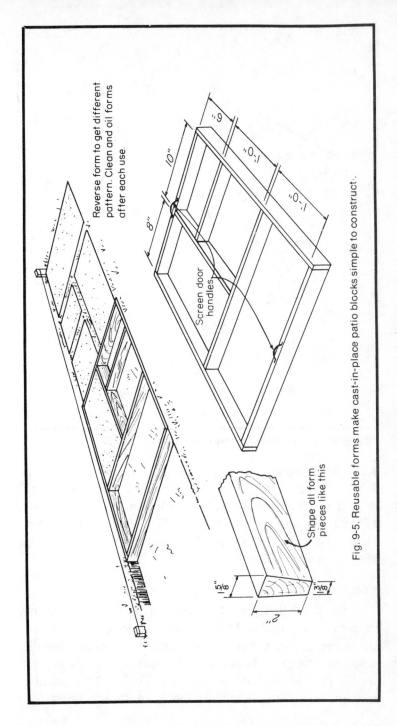

Reverse form to get different pattern. Clean and oil forms after each use.

Screen door handles

9"

10"

8"

1'-0"

1'-0"

Shape all form pieces like this

5/8"

3/8"

2"

Fig. 9-5. Reusable forms make cast-in-place patio blocks simple to construct.

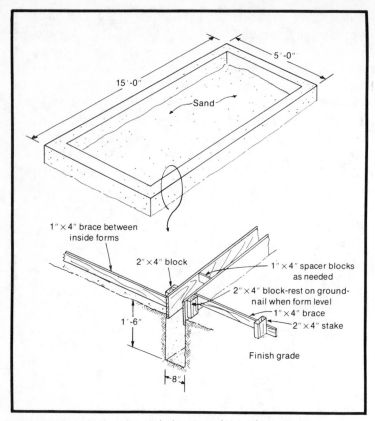

Fig. 9-6. Regular forming techniques can be used to make this sandbox.

aggregate. Gravel should be 1/2 inch maximum. The form is left in place after pouring until the concrete sets, then moved on to the next part of the walk. Finishing and curing are done as described in previous chapters. You can also cast other units in place, such as the sandbox shown in Fig. 9-6, by using regular forming techniques.

FORM CASTING

Almost anything can be made out of concrete with the proper forms or molds. The concrete lawn decorations you can buy along country roads are made with special molds, available from companies such as the Concrete Machinery Co., Drawer 99, Hickory, N. C. 28601. With a little ingenuity, you can make your own forming molds out of wood. Mailboxes, lampposts, and signposts are some of the things that can be constructed in this manner (Fig. 9-7 and 9-8). The mix here is the same as for casting in place. Forms are usually made of 1 × 6 or 2 × 6 lumber. The more deeply grained the lumber, the more interesting the posts will look. The same forms can be used for different types of posts, with only the length differing. Be sure to make your posts at least three feet longer than the desired height to allow plenty of room in the ground. If preferred, you can cast a base as shown in Fig. 9-7 for the mailbox. In this event, the post is made first, with reinforcing bars bent as shown. Nuts and washers are attached later. This will give firm anchorage to the base. You can also cast bolts into the bottom end of the post.

Reinforcing bars should be used in most posts to give them stability (unless you are casting a pipe integrally as shown in the lamppost, Fig. 9-8). The bars should be 1/4 inch (#2) or 3/8 inch (#3), depending on length and width. Wire mesh is usually used in curved objects. For beveled or rounded corners, insert moldings of your choice (inside corner for rounded, triangular for beveled) along the corners of the forms. To make connections between two posts, as in the signpost, insert bolts for later removal. Don't forget to imbed hooks and eyes into the concrete before it hardens (Fig. 9-9).

For quantity work such as fenceposts, use multiple forms like the one shown in Fig. 9-10. Insert bulkheads as shown, if you want some of the posts to be shorter than the others. For all castings, the concrete must be thoroughly spaded around form edges. Tap the forms lightly after pouring to release air bubbles.

If you want a certain uniform design at the top of any form, like the birdbath shown in Fig. 9-11, attach wood or metal templates to a central post. The templates are rotated to strike off the top as soon as the concrete sets firmly enough. To make a hollow in any form, like the base shown in Fig. 9-11, form a clay base and cover the top with oil. Since templates are tricky to use, this type of casting is done bottom-side up, with the "showing" side formed by the clay.

Flower boxes, garden benches, and almost anything else can be formed by using a little creativity. For some items,

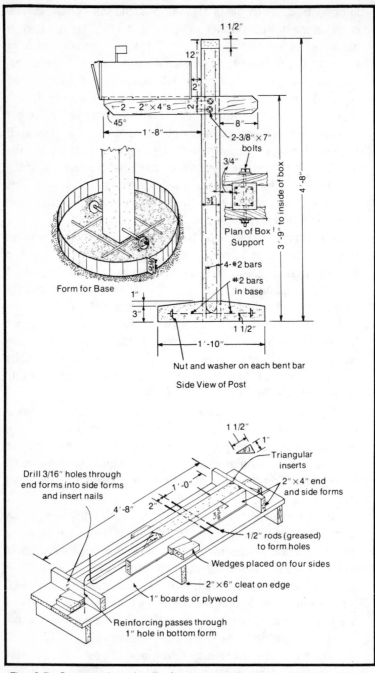

Fig. 9-7. Construction details for post-type projects, like a mailbox.

118

such as the rounded alternate bird-bath pedestal, the formwork can be rather intricate. In this case, the mold is made of 1/4 by 1/2 strips of wood alternating with half-round molding. The strips and moldings are nailed to form supports, one half at a time. Make sure that the moldings cover the strips slightly so that there are no openings. The form halves are then fastened together and concrete is poured as usual.

By using different type moldings, various edging and decorative effects can be achieved. The cove moldings in the standard bird-bath pedestal form give it classic lines, while the screen moldings used in the flower-box forms add a design effect to break up the monotony of the plain concrete sides (Fig. 9-12).

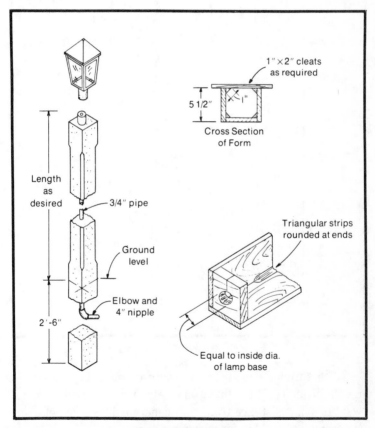

Fig. 9-8. Construction details for a lamppost.

119

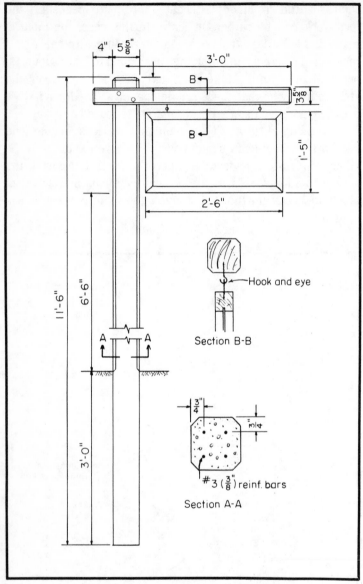

Fig. 9-9. Making a cast-concrete signpost.

The simpler designs shown here can be adapted without much difficulty. If the formwork gets more involved, you may want to ask assistance from your concrete dealer to make sure the forms are right before you pour them.

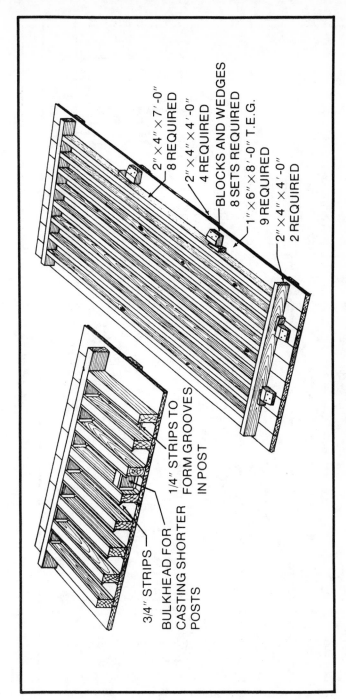

3/4" STRIPS

BULKHEAD FOR
CASTING SHORTER
POSTS

1/4" STRIPS TO
FORM GROOVES
IN POST

2" × 4" × 7'-0"
8 REQUIRED

2" × 4" × 4'-0"
4 REQUIRED

BLOCKS AND WEDGES
8 SETS REQUIRED

1" × 6" × 8'-0" T.E.G.
9 REQUIRED

2" × 4" × 4'-0"
2 REQUIRED

Fig. 9-10. Forms for multiple-post castings.

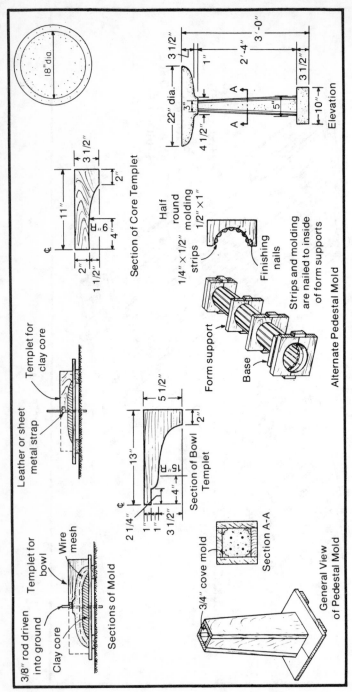

Fig. 9-11. How to make an ornamental birdbath.

18" dia

3'-0"
3 1/2"
1"
2'-4"
3 1/2"
22" dia.
3"
4 1/2"
5"
10"
Elevation
A
A

Section of Core Templet
3 1/2"
2"
11"
6"
4"
2"
1 1/2"
℄

Templet for
clay core
Leather or sheet
metal strap

Half
round
molding
1/2" × 1"
1/4" × 1/2"
strips
Finishing
nails
Form support
Base
Strips and molding
are nailed to inside
of form supports
Alternate Pedestal Mold

Section of Bowl Templet
5 1/2"
13"
2"
4"
℄
2 1/4"
1"
1"
3 1/2"

3/8" rod driven
into ground
Templet for
bowl
Wire
mesh
Clay core
Sections of Mold

3/4" cove mold
Section A-A

General View
of Pedestal Mold

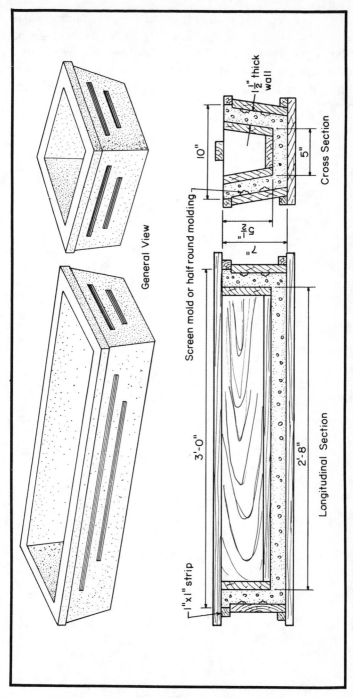

General View

1½" thick wall

10"

5"

Cross Section

Screen mold or half round molding

5½"

7"

3'-0"

2'-8"

Longitudinal Section

1"x1" strip

Fig. 9-12. Molding nailed to the sides of these forms will make the attractive grooves in the finished concrete.

123

To make decorative objects out of concrete, first make a plaster cast of your model. Give the model a few coats of shellac to prevent bonding. You may need two or more parts so that the plaster mold can be removed. Once the mold dries, remove it, reassemble the parts, and fill the inside with concrete.

Part 2 :
Concrete Masonry

10

Concrete Block, Tools, and Materials

Poured concrete is the method of choice for making slabs, stairs, and other easily-formed structures, but high walls are difficult to form and can be dangerous if improperly constructed. Concrete block is generally preferred for walls and buildings (Fig. 10-1). It is easy to lay block upon block and there should be no worries about the walls tumbling down if you do a decent job. Once the technique is learned, a reasonably skilled handyman can put up concrete block in surprisingly short order.

There are many projects that lend themselves to block instead of poured concrete. Some examples are fireplaces, barbecue grills, retaining walls, and walls of garages and homes (including additions). You can also build handsome fences and privacy walls with the many types of decorative masonry which are, in reality, glorified concrete blocks (Fig. 10-2). Many people still build their own in-ground swimming pools using concrete block, either with a vinyl liner or mortar coating (the block alone is too rough and porous). Patio blocks, illustrated in Fig. 10-3, are yet another form of concrete block.

In addition to glamour masonry, there are also special facings on solid block made to blend with other types of materials. *Stone-faced* block is designed to blend with the

127

Fig. 10-1. Standard concrete block is used, as is, for many farm and commercial buildings. It is also used with brick or other facing for homes.

stone foundations of older homes. *Slush* block has a rustic look for outdoor use. *Shadow wall* block has recessed corners which can create 1000 different wall patterns. It's ideal for basement recreation rooms (Fig. 10-4). Furthermore, most "brick" walls are really brick veneer, with concrete block providing the stability (see Chapter 17).

Early concrete blocks were solid units, cast in wooden forms such as soap boxes. They were molded by hand out of cement and gravel. If you think today's concrete block is

Fig. 10-2. Decorative blocks in many shapes and sizes make handsome screen walls or privacy fences.

Fig. 10-3. Patio block is ideal for outside, horizontal use.

heavy, you should have tried to lift those! The oldies were big and hard to handle. Late in the nineteenth century, some inventive fellow hit upon a very simple, but revolutionary idea—to put holes in the block. The "hollow" block was strong enough and considerably lighter.

It is uncertain who exactly made the first hollow block, but the first commercial manufacturing process was designed by Harold S. Palmer. In 1900, he patented a mold which produced the basic type of block which we use today. The molds were filled with cement and aggregate, then hand-tamped into individual molds. Two men could produce about 80 blocks in eight hours. More machines were invented to speed the process, and by 1924 automatic tamping machines were able to produce 3000 blocks a day. Now, modern vibration machinery is used instead of the tamping devices, and one machine is able to turn out 20,000 blocks per day. The entire process is fully automated and requires only a few workers.

Cinderblock is a type of masonry that uses waste cinders instead of the heavier aggregates. It was invented in 1913 to

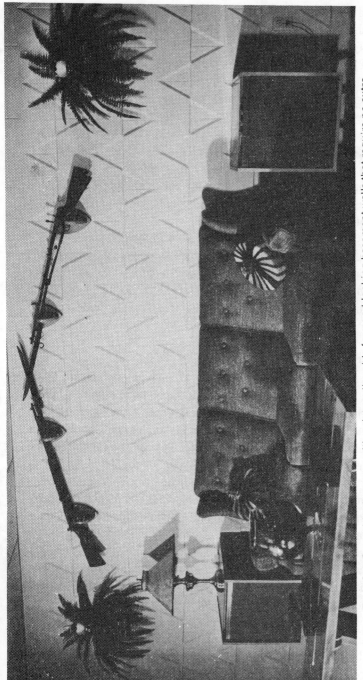

Fig. 10-4. Some block comes with a decorative side for good-looking basement walls that require no extra interior work (Courtesy National Concrete Masonry Ass'n.).

131

produce a type of block that was lightweight. Also, nails could be driven into it. Concrete block is more durable and better suited where strength is the major factor, but cinderblock is an excellent alternative for cases where it is more important to be lightweight and able to accept a nail without shattering. Other lightweight blocks may be made of pumice, slate, clay, or expanded materials such as shale, slag, or fly ash. They usually weigh approximately half as much as a regular block made of sand, gravel, crushed limestone, or blast-furnace slag. Since aggregates make up about 90 percent of the composition of the block (along with portland cement and water), the *type* of aggregate accounts for the differences in weight.

Other materials sometimes used in making block include air-entraining agents, accelerators to speed hardening, and coloring agents. High-early-strength cement is sometimes used to give the block greater strength during setting and curing. This helps to reduce damage during handling and delivery.

SIZES AND SHAPES

Although blocks were made in varying sizes at one time, virtually all block is now made in modules of a nominal eight inches. The typical *stretcher* block is nominally 16 inches long, eight inches wide, and eight inches deep. The actual size of the block is 3/8″ less on all sides to allow for the mortar joints. Thus, the stretcher block is really 15 5/8″ by 7 5/8″ by 7 5/8″. Partition block nominally four inches thick (sometimes six inches) is 3 5/8″ in reality (Fig. 10-5). This may seem to make planning more complicated, but actually the reverse is true. When laying out a wall, you should plan it in eight-inch modules, because the blocks will be laid up with the mortar in between, bringing the size up to 16 inches for the stretchers and eight inches for half-blocks or a sideways stretcher. The latter block is used for corners.

The stretcher blocks described above are used for the field area of your wall. They are made with either two or three holes or *cores* (Fig. 10-6), which are tapered so that one side is slightly larger than the other. In construction, the larger cores are laid down, so that there is more room on the top for mortar to be laid. The part with the smaller cores or larger block area is therefore the top. The section between the cores is called the *web*. The four larger faces are called *shells* and the smaller two, the *ends* (Fig. 10-7). Ends can either be flat or concave. Flat ends are used in corners or next to other walls. Concave ends are used in most field work. The *ears* that extend in the concave ends are called *flanges*.

In addition to the most common stretchers, other units in wide use include:

corner block—the block with one flat end, as described above. May have two flush ends.

header block—has a recess in which a lintel or similar unit is placed (see below).

lintel block—used to cross over an opening. It is shaped like a "U" or "W".

partition block—a thinner unit, usually four or six inches thick, for use in non-load-bearing walls.

L-corner block—used when walls are more or less than the usual eight inches, to make up the difference at corners.

These and other types of block are illustrated in Figure 10-5.

MORTAR COMPOSITION

Concrete mortar is composed of aggregate, portland cement, water and lime. The aggregate is made from sand, either natural or manufactured. Manufactured sand is made from crushed stone, gravel, or air-cooled blast-furnace slag. In either case, the particles are very fine.

Since the mortar joints must be as strong as the block itself, the proper mortar mix is important, just as it is in concrete. The do-it-yourselfer should use *masonry cement*, which has the lime premixed, and follow manufacturer's directions. The usual mix is one part masonry cement to three parts sand, with just enough water to make the mix workable. A mud-like consistency is the ideal. For smaller jobs, premixed mortar containing cement, lime, and sand is an even better bet. Simply add water according to instructions. To mix your own mortar, see Chapter 15.

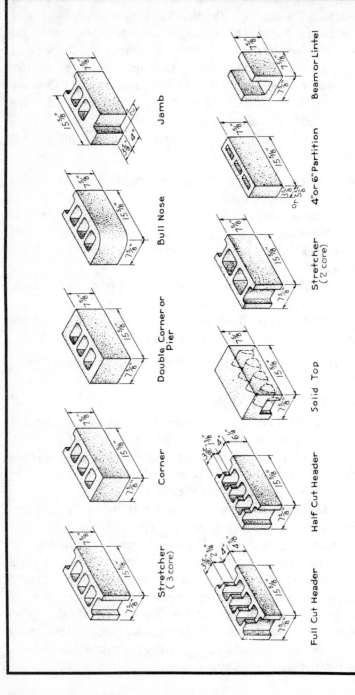

Jamb

Bull Nose

Double Corner or Pier

Corner

Stretcher (3 core)

Beam or Lintel

4" or 6" Partition

Stretcher (2 core)

Solid Top

Half Cut Header

Full Cut Header

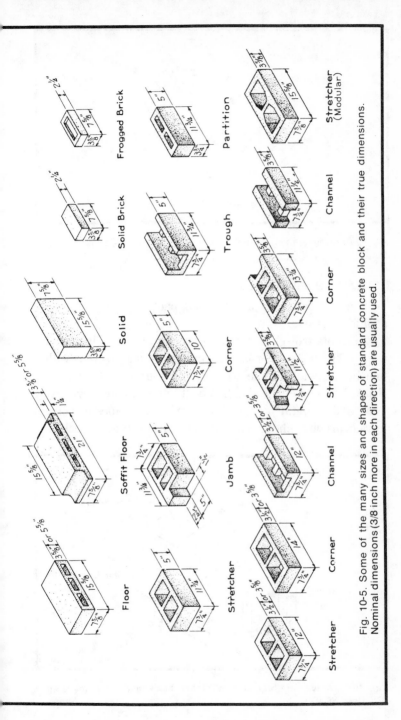

Fig. 10-5. Some of the many sizes and shapes of standard concrete block and their true dimensions. Nominal dimensions (3/8 inch more in each direction) are usually used.

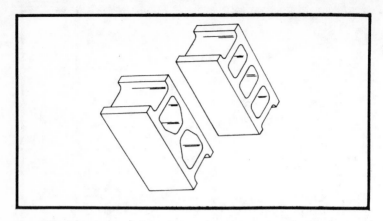

Fig. 10-6. Some standard block comes with three cores, some with only two.

TOOLS AND MATERIALS

For this type of work, a strong back is your best tool. In addition, however, you will need:

mason's trowel—the most important tool in all masonry work (Fig. 10-8). This is the common pointed trowel, with a thin blade, wide at the handle, and connected to it by the *shank*. The end of the shank is bent into a handle, covered by wood and a metal *ferrule*. There are many types of pointed trowels around; choose one carefully for this type of work. The

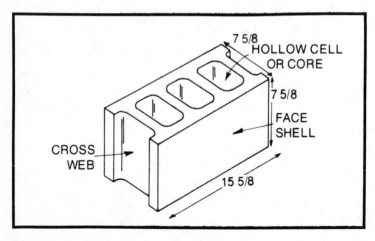

Fig. 10-7. A typical three-core stretcher block and its nomenclature (Courtesy NCMA).

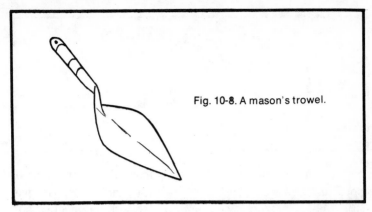

Fig. 10-8. A mason's trowel.

blade should be at a slight upward angle from the handle, and should be made of high quality flexible steel. Try it out for lightness and balance. Check the weld at the shank and the blade. It should be strong and smooth. Remember that your trowel will be subjected to a lot of friction between the block and the mortar. A cheap one will be clumsy and will wear out fast. A ten-inch length is best, but you may find a shorter one more comfortable.

brick hammer—actually a mason's hammer, but more commonly called a brick hammer (Fig. 10-9). It has one square end, and a long, tapered chisel at the other end. The square part is used like an ordinary hammer, the chisel *peen* is for cutting off pieces of block and dressing edges.

bolster—also known as a brick chisel, blocking chisel, or a *set*. It is used to cut brick and concrete block (Fig. 10-10). One edge of the wide blade is beveled. The beveled edge always faces toward the user.

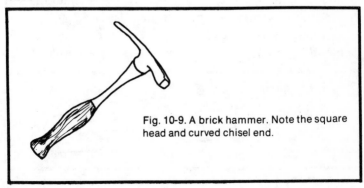

Fig. 10-9. A brick hammer. Note the square head and curved chisel end.

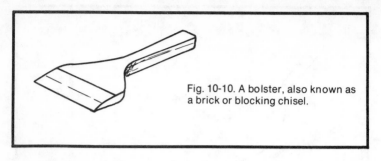

Fig. 10-10. A bolster, also known as a brick or blocking chisel.

mason's level—used like an ordinary level, but made of wood and considerably longer than the standard spirit level (Fig. 10-11). Each block laid up should be leveled and plumbed, and this tool accomplishes both functions. For concrete block work the level should be 42 to 48 inches long. Practically all of these levels have vertical vials for plumbing, but make sure the one you want to buy has this, just in case.

folding rule—the standard six-foot rule is all right here, but a mason's *spacing rule* is even better. In addition to the standard markings on one side, the reverse has a series of marks representing course heights, plus various combinations of unit size and joint thicknesses. This feature is not necessary for standard block, but is handy for decorative block and brickwork.

steel square—a small metal square for checking corners and for making accurate cuts in block. The large carpenter's squares are too bulky for this type of work.

mason's line and pins—a good line with plenty of stretch and give, used for keeping your wall straight, level, and plumb. It is attached to pins which are driven into the soft mortar at the corners. A chalk line may also be helpful.

jointers—special Z-shaped tools for dressing mortar joints. Round tools make concave joints, square-ended types

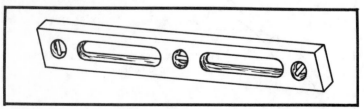

Fig. 10-11. A mason's level, made of wood 42 to 48 inches long.

are for V-joints. Other varieties are available, too. A piece of 1/2-inch copper tubing does almost as well for concave joints.

mortar box and hoe—used for mixing mortar. Professionals often use a power mortar mixer, but most handymen will do nicely with handmixing.

mortarboard—used to carry the mortar from the mortar box to the work site (Fig. 10-12). You can buy these made of wood or steel, equipped with a handle, but a piece of exterior plywood should do just as well to begin with. Screw a handle to it for easier carrying.

story pole—a 1 × 2-inch board, about 10 feet (a lone *story*) high, marked to check the height of each course including mortar (Fig. 10-13). In addition to the basics—block and mortar—there are several other materials you may have to use:

metal ties—used when building two or more *wythes* or tiers of masonry, and also, when joining intersecting walls. They are made of 3/16″ copper or zinc-coated steel and are usually shaped like a "Z" or a square (Fig. 10-14). One is used for every four square feet of wall. When tying concrete block backup to brick veneer, corrugated metal or dovetail ties are often used (Fig. 10-15). One is installed for every two square feet of wall.

reinforcing bars—similar to the reinforcing rods used in some concrete work. Plans usually refer to them as *re-bars* and specify the size in eighths of an inch. A #6 re-bar would be 6/8 inches, #8 equals 8/8 (or one) inch, etc. Re-bars are usually

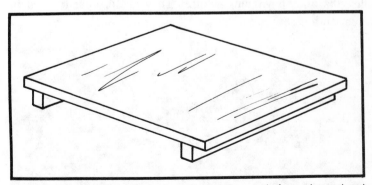

Fig. 10-12. A mortarboard may be purchased or made from plywood and scrap lumber.

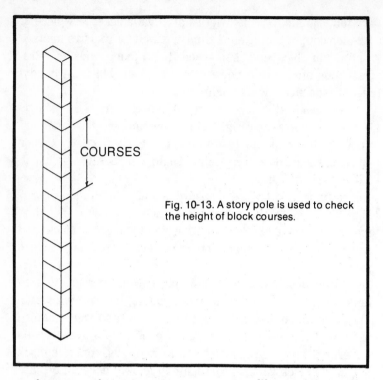

COURSES

Fig. 10-13. A story pole is used to check the height of block courses.

used to strengthen corners or stress areas like retaining walls. They are installed vertically and anchored in the footings.

joint reinforcement—used between courses for strengthening (Fig. 10-16). They are made of steel and are usually shaped like a *truss*, or ladder, with two bars on each side and cross rods between. These rods come in 10-foot lengths and in varying widths. They are ordinarily used in every other course.

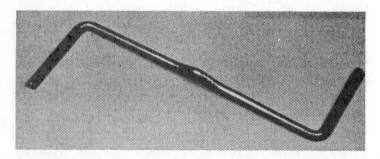

Fig. 10-14. A metal "Z" tie is used to connect double walls or intersections.

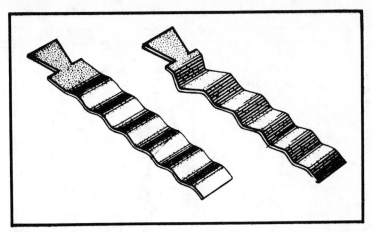

Fig. 10-15. Corrugated ties are used to connect brick veneer and backerblock (Courtesy NCMA).

anchors—several types are in use. An anchor bolt is embedded in the core or between tiers or *wythes* so that wood or other materials can be attached. *Strap anchors* are long metal ties bent at each end. They are used to furnish a rigid connection between bearing walls. Regular ties or hardware cloth can be used for less rigid anchorage.

joint fillers—similar to the molded rubber used for isolation joints in concrete work. They are used for control and expansion joints, generally vertically.

Fig. 10-16. Joint reinforcement. At left is the ladder type, at right is the truss type.

Fig. 10-17. Running bond is recommended for most projects, especially for beginners.

grout—a thinned mortar used for filling cores or other empty spaces in block walls. Ordinarily, use leftover mortar, adding water until the mixture pours easily.

BOND

Bond refers to the pattern in which concrete block or brick is laid. By far the most common is the *running* bond pattern which uses 8 × 8 × 16 block. In running bond, one course is laid over the other, so that the joint is directly in the center of the block below (see Fig. 10-17). The beginner should stick to running bond and standard-size block unless there is good reason to do otherwise.

11

Construction Techniques

All concrete masonry jobs require planning. You must be especially exacting when working with block. Depending on the needs of the project, it is wise to plan every dimension in multiples of eight inches. Use 16 inches for field block, counting the mortar, but subtract 3/8″ at each end because there will be no mortar there. Half-blocks (eight-inch cubes) are used when 16 inches is too long. The closer you can get your project to eight-inch modules, the simpler and cheaper it will be.

FOOTINGS

Block is always laid on a poured concrete *footing*. The footing must be at least two feet below the surface or six inches below the frostline. Footings should be twice as wide as the wall, so the standard footing is 16 inches. The usual footing is also as deep as the wall is wide, in this case eight inches.

For extra strength, key the footing with a groove down the center, so that the mortar fills in and creates a firmer bond between the footing and the first course. A tapered, 1 × 4 board oiled and embedded in the footing will accomplish this purpose.

No special finishing is necessary for footings, since no one will see them, but they must be as level as possible to provide a

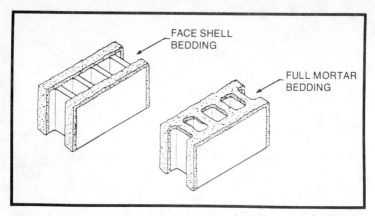

Fig. 11-1. Face-shell bedding is used for most mortar joints. The first course and other special usages require a full mortar bedding (Courtesy NCMA).

good start for your block. Retaining walls or other walls that will undergo a lot of stress should be braced with steel reinforcing rods through the block cores. If rods are used, they should be embedded in the footings for maximum strength. Be sure to align the rods so that they will go through the cores of the block.

STARTING

To avoid fitting problems, it is a good idea to make a dry run by setting the blocks on the footing with a 3/8″ space for mortar in between each block. Snap a chalk line in front for positioning. Mark each end and the mortar spaces. Later as you get more proficient, you may just make marks at the ends.

For the first course, you will use a *full bed* of mortar. Other courses will require *face-shell bedding.* A full mortar bed simply means that the entire horizontal surface is covered with mortar instead of just the front and back (Fig. 11-1). Footings should be wet down first to minimize absorption. *Never* wet the block itself.

Always start at the corners and work toward the middle (Fig. 11-2). The mortar bed should be laid down the center of the footing and furrowed down the middle with a pointed trowel, allowing the face shells of the block to absorb most of the mortar. Each block must be checked for level and plumb.

144

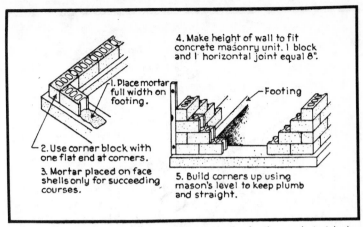

4. Make height of wall to fit concrete masonry unit. 1 block and 1 horizontal joint equal 8".

1. Place mortar full width on footing.

— Footing

2. Use corner block with one flat end at corners.

3. Mortar placed on face shells only for succeeding courses.

5. Build corners up using mason's level to keep plumb and straight.

Fig. 11-2. Lay down a full bed of mortar on the footing and start laying block at each corner.

Doublecheck with the mason's level every three or four blocks. Make sure that the mortar bed is approximately 3/8-inch at every joint and that the blocks are all aligned correctly. Tap the block with your trowel handle for minor adjustments. Work carefully but not too slowly, as all adjustments must be made before the mortar stiffens. And make sure you lay the block with the smaller ends of the core holes up in order to provide more space for the mortar.

The vertical joints between each block should contain mortar at each edge of the block. Set the blocks on end before placing, and *butter* the other end with mortar (Fig. 11-3). Start

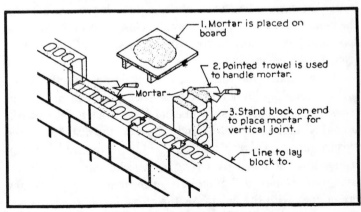

1. Mortar is placed on board

2. Pointed trowel is used to handle mortar.

Mortar

3. Stand block on end to place mortar for vertical joint.

Line to lay block to.

Fig. 11-3. Steps to professional blocklaying.

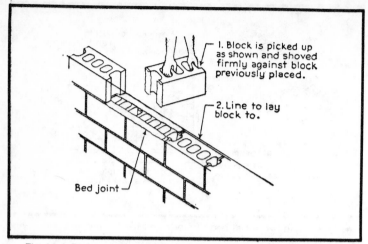

Fig. 11-4. Pick up buttered block and place against previous block.

with one and do two or three blocks at a time as you get more proficient. Press the mortar down onto each edge so that it will not fall off, then lift the block into place (Fig. 11-4). Press down firmly and push against the previous block. If you have used enough mortar, it should ooze out of all the joints. Cut off the excess mortar with the trowel and return it to the mortarboard. If the mortar gets too stiff, *retemper* it by adding water. Throw away unused mortar that was mixed more than 2 1/2 hours previously.

You should have planned your wall so that the last *closure* block will be 16 inches counting mortar. If this is the case, butter the *ears* on both sides of the last block, plus the edges of the blocks already in place. Lay the closure block in from the top, being as careful as you can not to knock off the mortar from any of the edges.

If the space for the closure block is not exactly 16 3/8 inches (counting the joints on both sides), you will have to cut block. Professionals use a masonry saw, but amateurs will probably have to do it by hand. Mark off the distance on the closure block, (subtracting 3/4 inch for the mortar) score on both sides, and cut it with your bolster. For an accurate cut, use your square to make an even mark. Chip off any projections with the bolster or the chisel end of the hammer. Don't worry about minor gouges. They will be filled in with mortar. If the cut is too far off the mark, discard the block and cut another.

After the first course is in place, lay succeeding courses up from the corners. It is particularly important that the corner blocks be accurate. Block is almost always laid in a *running bond*, in which the top block overlaps two bottom ones equally. Lay three or four courses in each corner before joining the bottom one. Check and double check with the story pole for proper height (Fig. 11-5). Use the level to check for horizontal plumb. Using the story pole diagonally across the joints helps check horizontal spacing. String a line from one corner to another to insure accuracy, using line pins set into the mortar. Make sure the line is taut.

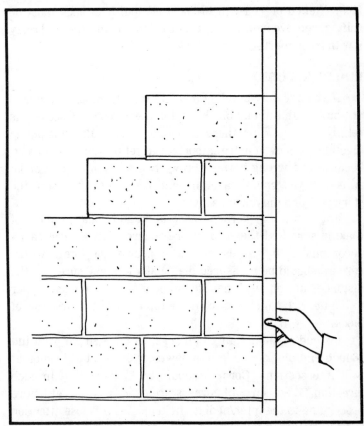

Fig. 11-5. Use the story pole to make sure that each row is the proper height for block plug mortar.

After the first course, face-shell bedding is used. Mortar is spread along the front and back of each row instead of over the entire surface. Except for this difference, the succeeding rows are placed exactly as the first one was placed. Align the top edge of each block with the bottom of string. Stay about 1/16 inch inside the line so you don't crowd it and cause bulging of the wall.

As you work, be careful not to spread the mortar too far ahead. Two or three blocks ahead is plenty for the beginner. If it starts to harden, mix fresh mortar. By the same token, doublecheck your work as you proceed, filling in any open joints or holes in the mortar while it is still plastic. Do not attempt to change the position of any block after the mortar has stiffened or you will break the bond. If a block must be shifted, remove it entirely, scrape off the old mortar, and relay it with fresh mortar.

BUILDING CORNERS

If you are only building a single wall, the ends of the wall are built with full end blocks and half end blocks. Checking is relatively easy. When there are two or more walls being put up together, it is even more important to get the corners correct, because one wrong corner throws two walls off kilter. Make sure that you check carefully and frequently with both the story pole and the mason's level.

For a two-way corner, there is an additional problem of making sure that the corners are square. This is most easily accomplished by the use of batter boards and lines. All the corners should be measured diagonally to make sure that the distances are equal. Lines should be strung from each corner to the next, 1/4 inch out from the top front of the first row of block.

Instead of using half end blocks, as for a single wall, full 16-inch end blocks are laid in sawtooth or dovetail fashion at wall intersections. Corners should be built up first in each direction. Make sure all corners are exactly on target before proceeding to the intermediate field block. Pay close attention to your plan to make sure that you have allowed openings for doors and windows.

All joints should be tooled with the jointing tool as soon as they are thumbprint hard. Concave joints are best, but a V-shaped joint is also watertight (Fig. 11-6).

CONTROL JOINTS

Most do-it-yourself jobs do not require control joints, but they may be necessary where forces acting on the wall might cause cracking. Retaining walls may require them, for example. You should really get professional advice on this if you have any doubts.

A control joint is a continuous vertical joint instead of the usual staggered one. Eight-inch half-blocks are used on every other course. The easiest way to make a good control joint is to run building paper or roofing felt down along the space between the *ears* of one side of block, and fill the rest of the space with poured concrete. This provides support, but allows some movement because the concrete isn't bonded on the building paper side. Special tongue-and-groove, or preformed gasket control joint blocks are available in some areas (Fig. 11-7). To accomplish the same purpose, Z-bar ties can also be used. Exterior joints should be raked 3/4″ deep and filled with caulking compound.

REINFORCEMENT

The strongest type of reinforcement is the use of vertical bars set into the footing. This should be used where there is a lot of pressure, such as in a retaining wall. All cores containing reinforcing bars are filled with grout later. To strengthen regular walls, horizontal reinforcing bars are placed between every other course. Be sure that the steel is fully embedded in the mortar bed. There should be at least 5/8″ mortar cover in the front and 1/2″ at the rear. Where bars meet, overlap them at least six inches and weld.

Any wall which is subject to water pressure behind it, such as a retaining wall, should contain *weep holes* to allow the water to escape. If water is allowed to build up behind it, the wall may fail. Weep holes are constructed by leaving spaces in the mortar joints of the first row of block, or by installing pipes at a downward angle.

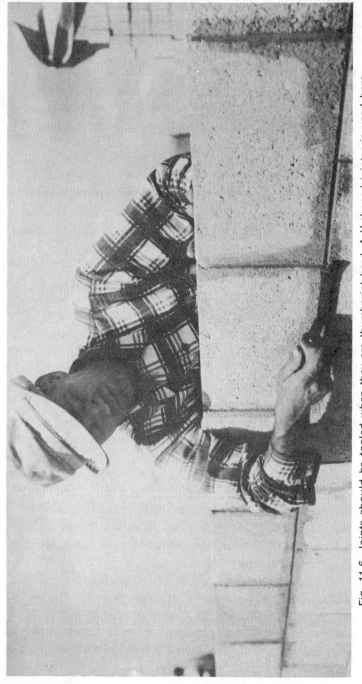

Fig. 11-6. Joints should be tooled when they are thumbprint hard. A V-shaped jointer is used here.

COMPOSITE WALLS

Most brick walls are not constructed of solid brick; they use concrete block or wood framing as a backer. These are called *composite walls* in the trade, and *brick veneer walls* by architects and real estate people. (See Chapter 17 for more on this.) Most stone walls are made the same way, and some exterior wooden walls have block backup, too.

Backerblock is put up the same way as other concrete block, except that some provision must be made for bonding the block to the brick or stone. An experienced mason often uses either the block or brick itself to bond the two elements, but the easiest way for the handyman to secure the walls together is with "Z" or rectangular metal ties. The ties are inserted into the mortar every 16 inches vertically and every 36 inches horizontally. Corrugated ties can also be attached with concrete nails. (Many building codes require even more frequent bonding, usually every one tie for every two square feet of wall area. This generally is translated into one tie every two feet both horizontally and vertically.) The other end of the tie is embedded in the mortar of the veneer wall (Fig. 11-8).

For watertight composite wall construction, the block wall should be *parged*. This simply means that a coat of mortar is

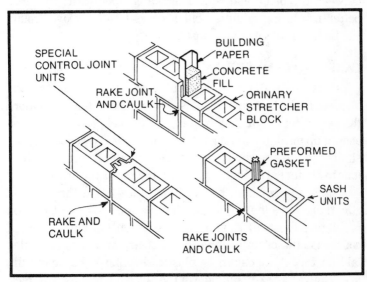

Fig. 11-7. Ways to form control joints if needed (Courtesy NCMA).

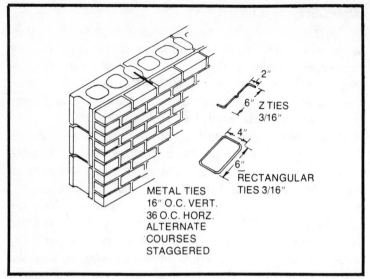

Fig. 11-8. A composite block and brick wall is tied together with "Z" or rectangular ties staggered as indicated (Courtesy NCMA).

applied to the side of the block that is next to the brick or stone. To avoid pushing the block out of line, hold the mason's level on the other side while the mortar is applied. Foundation block should be parged outside to insure a watertight wall. Cavities between the block and veneer are later filled with grout. For better insulating qualities blank cores located above-ground can be filled with granular insulation (Fig. 11-9).

CARE AND CLEANUP

Exterior walls are vulnerable to high winds and the elements. The beginner is advised to avoid building walls more than eight feet high. Any wall higher than eight feet should be shored up until the mortar hardens. Temporary wooded braces similar to the one shown in Fig. 11-10 should be erected every four to six feet. The higher the wall, the closer the braces.

Since the cores in block allow water to get through and stay there for a long time, causing moisture and condensation problems, the wall should be covered on top until the final course is laid, or until some sort of coping or other material is laid on top. A tarpaulin or plastic sheeting is laid carefully over the work, extending down the sides for at least two feet.

Spare block or other heavy material should weigh the tarp down so that it won't blow away.

When the wall is completed, holes left by line pins and nails should be filled with mortar. Take care at this time and during regular construction not to smear mortar on the surface of the block. Mortar stains are hard to remove. Any mortar that drops and sticks on the wall should be allowed to harden. If you try to remove it while still wet, it will smear. After it is dry, scrape off the hardened mortar with a trowel or putty knife. Smaller smudges and other drippings can be removed by rubbing the surface with a small piece of broken block. Any smudges still left can usually be eradicated by going over the area with a fiber brush.

Paint will brighten up plain concrete masonry, but be sure that the surface is scrubbed clean. Alkali in both block and

Fig. 11-9. Insulation is vital in these energy-short times. To make concrete block even more effective, fill all cores that are not grouted with poured insulation.

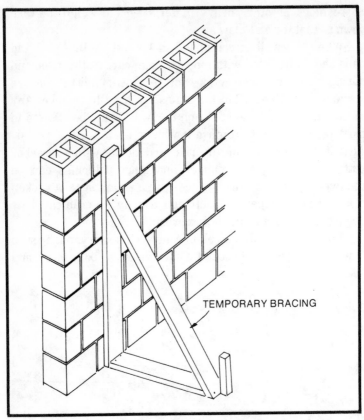

Fig. 11-10. Walls higher than eight feet should be shored up with temporary braces until mortar hardens (Courtesy NCMA).

mortar is destructive of many paints, so wait at least a few weeks before painting. And be sure to ask your paint dealer for alkali-resistant paint that will adhere well to the surface.

12

Specialized Concrete Masonry Techniques

The experienced mason can do almost anything with concrete block, but the beginner should shy away from the more difficult tasks. Basically, however, even the most complicated construction is a variation of the wall-building techniques previously described. There are several special applications, which the handyman can learn, that are useful in some of the more complicated situations.

BUILDING LINTELS

Doors and windows are formed by openings in block construction. Framing them is a job for the carpenter. The problem for the mason—not a difficult problem, really—is bridging the gap over the opening. There should be no special techniques involved in forming the sill below the window.

Good planning will facilitate construction of door and window openings. As illustrated in Fig. 12-1B the openings can be simply constructed with whole and half block, or they can be a headache if you've provided an opening in a difficult place, as in Fig. 12-1A. Where possible—and it usually is possible—plan windows and doors to make it easy on yourself.

Lintels are horizontal crosspieces placed over doors or windows, and they come in several types. The simplest type is a precast concrete lintel which is easily set in place over the

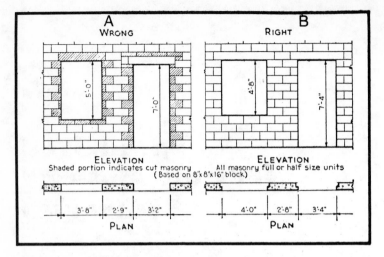

Fig. 12-1. Construction is infinitely easier if it is planned so that whole or half blocks are used instead of cutting a lot of block.

opening like any other block component. The concrete should be reinforced for greatest strength.

Steel can also be used as a lintel; "H" or "I" beams can be used alone, or a steel angle iron can be used as a bearing for block. An interior lintel using an angle iron and partition block is shown in Fig. 12-2. There are also special masonry units which can be used to build a block lintel (Fig. 12-3). These

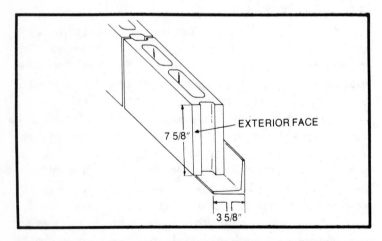

Fig. 12-2. Steel angle irons can be used with standard block (in this case partition block) to form a strong lintel.

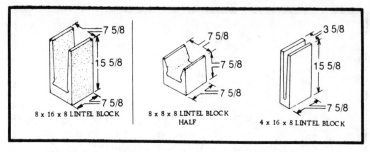

Fig. 12-3. Some of the most commonly used lintel blocks.

units are supported by temporary shoring (see Fig. 12-4), and are strengthened by reinforcing bars and poured concrete fill. Non-corroding metal plates should be installed under the bearing ends of each lintel to enable it to slip. Control joints at the ends of the lintel also allow lintel movement without bringing down the entire wall.

SILLS AND COPING

Sills and *coping* (the top layer of a masonry wall) can be made of wood, stone, or other materials. Precast masonry sills and coping units are installed after the wall has been built. These must be carefully mortared to prevent leakage into the wall below. These units should be placed in a full mortar bed with the ends tightly filled with mortar or a caulking compound.

Coping and sill material comes in many sizes and shapes. Whichever you choose, make sure that it has an overhang wherever exposed to weather (Fig. 12-5) and that all vertical

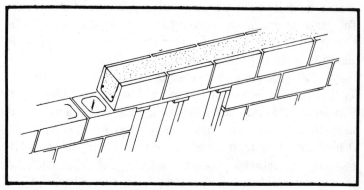

Fig. 12-4. A completed lintel using lintel block and temporary shoring.

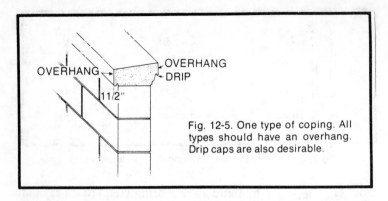

Fig. 12-5. One type of coping. All types should have an overhang. Drip caps are also desirable.

joints within the material itself are caulked tightly. Flashing installed with the coping must be carefully placed to prevent leaks.

PIERS AND PILASTERS

Piers and *pilasters*, which are supporting columns, are built similarly and have similar functions. The difference is that a pier is free-standing while a pilaster is actually part of the wall. Both can serve to support extra loads such as a beam or girder. For greater stability, cores of piers and pilasters are often filled with mortar and sometimes reinforcing rods.

Piers are usually built two blocks wide. The courses are alternated so that the shells face a different direction every other row. Pilasters are built the same way, except that they are built as an integral part of the wall, with every other course bonded to the wall itself. Non-corroding bearing plates are set in a full bed of mortar when used to support other work. The plate is set in the center to distribute the weight. The mortar should be spread heavier than the final thickness to allow the beam to be tapped down to proper height.

A pilaster gives added strength to the wall. It can be built for that purpose alone, or can be used for support like a pier. Sometimes they are used for both. When used primarily to strengthen the wall, they are usually placed every sixteen feet. Depending on structural needs, the pilaster should be 1/10 as wide as the distance between pilasters. The depth from the back of the wall to the back of the pilaster should be about 1/12 of the wall height. This usually adds up to two blocks,

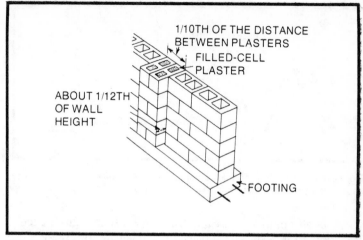

Fig. 12-6. A pilaster fits right into the wall. Piers are separate units, but are built much the same way.

alternated as discussed—but check the dimensions to make sure. Before footings are poured, make sure that they are wide enough to support the pilasters as shown in Fig. 12-6.

BOND BEAMS

A *bond beam* is often used in structures more than one story high, and can be used in one-story buildings to provide a continuous solid belt around the walls of the building. It serves not only to tie the wall together, but to prevent cracking and to serve as a solid bearing surface for floor joints and beams (Fig. 12-7).

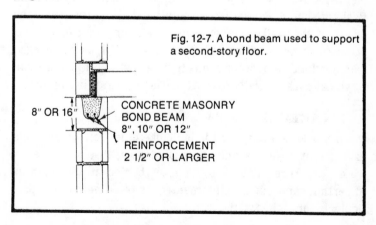

Fig. 12-7. A bond beam used to support a second-story floor.

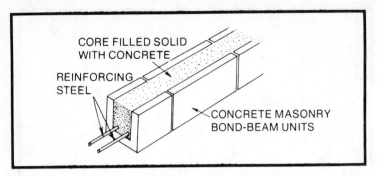

Fig. 12-8. Construction of a bond beam using specialized block. (You may improvise by knocking out the webs of standard block).

A bond beam is built much like a masonry lintel, except that the block below eliminates the need for the temporary shoring used for lintels (Fig. 12-8). Special bond beam or lintel units are used and are filled with re-bars and concrete. Regular units can be used, however, by knocking out part of the cross webs or face shells.

PARAPETS

Parapets are small walls which extend above the roof of a building. They may be built of any material, including concrete block, and can be any of the types described previously—single-wythe, composite, veneer, etc.

Because of the constant exposure of all sides of a parapet, however, extra measures must be taken to protect it from the elements. Full mortar joints are used throughout. Coping and flashing should also be laid in a full bed of mortar. Since the inside of the wall may serve as a retaining wall for excess water after heavy rains, the inside of the wall is often parged with portland cement for protection against water penetration. A typical parapet wall cross section is illustrated in Fig. 12-9.

BLOCK RETAINING WALL

A block retaining wall is built like other reinforced walls except that drainage must be provided because of the heavy water pressure that can build up behind it. Strength is important, and extra reinforcement is often necessary, particularly for higher walls.

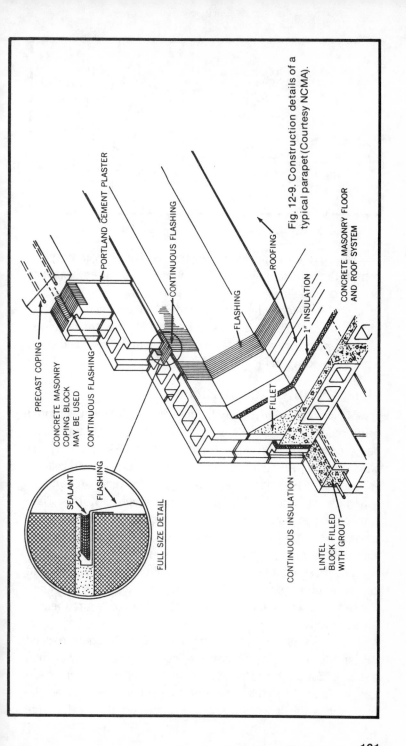

PRECAST COPING

CONCRETE MASONRY
COPING BLOCK
MAY BE USED

CONTINUOUS FLASHING

PORTLAND CEMENT PLASTER

CONTINUOUS FLASHING

FLASHING

CONTINUOUS FLASHING

ROOFING

Fig. 12-9. Construction details of a typical parapet (Courtesy NCMA).

1" INSULATION

FILLET

CONCRETE MASONRY FLOOR AND ROOF SYSTEM

SEALANT

FLASHING

FULL SIZE DETAIL

CONTINUOUS INSULATION

LINTEL BLOCK FILLED WITH GROUT

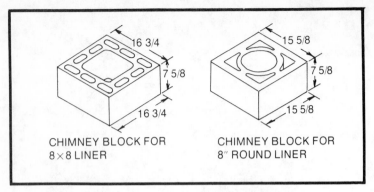

16 3/4

7 5/8

16 3/4

CHIMNEY BLOCK FOR
8×8 LINER

15 5/8

7 5/8

15 5/8

CHIMNEY BLOCK FOR
8" ROUND LINER

Fig. 12-10. Two of the more common types of chimney block.

To provide the necessary drainage, *drain tile* is installed behind the wall and surrounded by gravel, or *weep holes* are provided in the lowest course. Drain tiles should be connected to an outlet. Weep holes can be made by placing non-corroding pipe through the walls or by leaving holes in the mortar joints. Parging helps prevent water penetration in the rear and gives a better-looking surface in front.

CHIMNEYS OF CONCRETE BLOCK

Masonry chimneys may be built from solid block, grouted reinforced hollow block, or from chimney block especially designed for the purpose. Figure 12-10 shows some typical chimney block, manufactured so that a flue lining will readily fit inside. Flue liners are designed to function without softening or cracking at temperatures as high as 1,800 F. They protect the masonry from high temperatures and corrosive gasses. A single flue chimney of concrete brick, as shown in Fig. 12-11 is the type usually found in homes.

The first course is laid out as illustrated. The second course is laid up and the frame for the cleanout door is set. These frames are usually fitted with anchors to secure them to the masonry.

The work is continued to the level of the *headers* marked "H" in Fig. 12-11. The headers extend into the chimney to form a support for the first flue liner. Two or three more courses are laid to firmly fix the headers in position. Then, after the mortar has hardened sufficiently to support it, the first flue lining is set on the headers. The flue liner, which has a hole in the bottom, is then plumbed and centered in the chimney. The work is continued up to the top of the first liner, with the brick cut as required for the smoke pipe.

All flue liners are connected with mortar joints. Any projecting mortar is cut off from the inside to assure a good draft in the chimney. The process of laying up concrete brick around the flue lining, and setting the next lining, is repeated until the chimney reaches the required height.

At the top, the chimney is built up with cement mortar sloping from the flue liner to the edge of the masonry to provide drainage. (See the two-flue chimney in Fig. 12-12.) Cement mortar made from fine sand and portland cement is

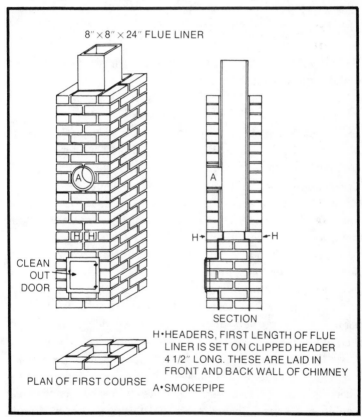

8" × 8" × 24" FLUE LINER

A

CLEAN OUT DOOR

SECTION

H•HEADERS, FIRST LENGTH OF FLUE LINER IS SET ON CLIPPED HEADER 4 1/2" LONG. THESE ARE LAID IN FRONT AND BACK WALL OF CHIMNEY

PLAN OF FIRST COURSE

A•SMOKEPIPE

Fig. 12-11. A single-flue chimney unit built from concrete block. Note the construction around the clean-out door and headers to support the flue liners.

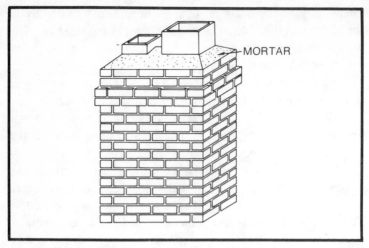

Fig. 12-12. A double-flue chimney. Arrow points to mortar cap sloped away from flue to provide good drainage (Courtesy NCMA).

floated with a wooden float to close any shrinkage cracks and to thoroughly compact the surface.

FIREPLACES OF CONCRETE BLOCK

A concrete masonry fireplace and chimney combination is built like the brick version detailed in Chapter 18. The combustion chamber is lined with fire brick using thin joints of fireclay mortar. The back of the combustion chamber slopes toward the opening to throw the flame forward. The sides of the chamber are angled or splayed to permit maximum radiation of heat into the room. The back and end walls of the fireplaces are made from reinforced hollow concrete blocks, and they support the chimney above.

A metal damper is installed at the top of the sloping back to stop drafts and heat loss from the room when the fireplace is not in use. The curved surface above the sloping back is called the *smoke shelf*. It acts as a barrier against cold air coming down the chimney, and deflects it upward into the smoke chamber. It directs the smoke to the bottom of the chimney, where the flue liners begin. When the chimney is constructed outside the house, the chimney should be anchored. Metal strips attached to the wood frame are embedded in the mortar, with the top of the chimney anchored through a bond beam.

A QUICK AND EASY BUILDING SYSTEM

If you'd like to put up a small structure in a hurry and without too much fuss, you should look into a new system which uses surface bonding cement instead of standard mortar. With this type of construction, block is simply laid up in its proper sequence using all the techniques and safeguards as discussed in this section. The difference is that mortar is not used.

When all the block is placed as desired, surface bonding cement such as Bonsal's Surewall is troweled over the entire surface on both sides (Fig. 12-13). After setting for the time recommended by the manufacturer, the wall is sturdy and strong. Openings are framed temporarily to hold the block until the cement sets. Check building codes before starting.

PATCHWORK AND REPAIR

No matter how good the original concrete or masonry job, there may come a time when repair or patchwork is necessary. A good concrete job, properly installed and left unabused, will easily last a lifetime. But there are trees with strong roots, children with hammers, and frosts. Concrete *can* be broken, heaved, cracked, and pitted—even if perfectly laid and cured. Imperfectly laid concrete (not your job, of course) can spall and flake.

The same holds true for the block that is made from concrete. Stone can be broken, as can brick. But most masonry problems are caused by poor or ancient mortar. If the mortar has badly deteriorated, the only recourse is to tear down the wall or structure and start afresh. Ordinarily, however, *repointing* or *tuck-pointing* (removing old mortar and replacing with new) is all that is necessary.

Concrete Repair

The art of concrete repair has advanced greatly in the past decade. New repair mixes, containing a vinyl latex or epoxy ingredient, have made small repairs fast and easy. Standard concrete mixtures do not take well to patchwork. You can't just take a standard concrete mix and slap it onto a broken corner. It won't adhere to the old work. Similarly, a low piece

A

B

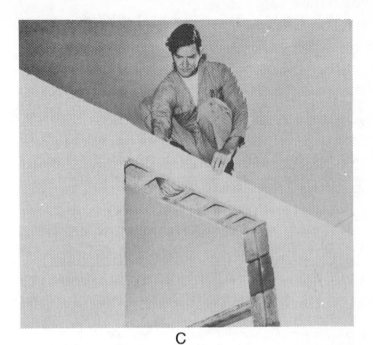

C

Fig. 12-13. Photo sequence shows construction of one-story building using surface bonding cement. Block is simply laid up without mortar, using the same techniques otherwise detailed. Surface bond is then troweled on. When dry, it can support a man's weight on one course as shown (Courtesy Bonsal Surewall).

of sidewalk cannot be built up with more concrete unless it is first *scarified* (roughened). Then room must be made for at least two inches of new work.

The more modern patching compounds can be *feathered*, or smoothed out to an imperceptible edge. They will adhere to old concrete because of the extra ingredients. But they are also very expensive compared to regular concrete.

The cure depends on the cause, and determining the cause is the first order of business. If a section of concrete has heaved right next to a large tree, you can be pretty sure that expanding roots are the culprit. It will do no good to patch up the offending piece, because it will undoubtedly be higher than the rest even before patching, and the same thing will happen as the tree continues to grow. If you want a nice, level walk, the only thing to do is to break up the old section with a sledge hammer, cut off the roots underneath, and lay down a whole new slab. If the tree is big enough to heave concrete, it will withstand having a few roots lopped off.

If that seems like too much work, an alternative (temporary, however) is to use patching cement to create a slope around the heaved portion. This won't solve the problem, but it will at least prevent people from tripping over the broken, raised portion.

When a concrete section is slightly lower than the adjacent ones, the cause may be some sort of earth undermining. Usually, however, this is a case of uneven settling due to poor preparation of the subgrade, or just freakish expansion and contraction. In either event, the easiest cure is to lay down enough patching cement to bring the surface level with the other part of the walk. If the sunken area is recessed, more than a half inch or so, this process may be too costly. A new topping may be in order.

Pitting, small cracks, broken step corners, and other minor damage can usually best be cured by using special patching cement. Large cracks can be fixed by undercutting the entire break with a cold chisel. The bottom of the man-made crack should be wider than the top, so that the new concrete will not come out easily. (Your dentist does the same thing with a filling.) With this method, you can use regular concrete. Bagged concrete mixes are ideal for this. Since the patch will usually be darker than the rest of the concrete, you should add a little white portland cement to the mix.

Small scaled or spalled areas may be covered with special patching cement, but if your entire concrete surface is deteriorating, you know that something is radically wrong. The concrete may have been improperly mixed or cured, or perhaps you've been using chemical deicers containing ammonium nitrate or ammonium sulfate. Vow never to use them again, and resign yourself to a rather extensive job. You can hardly afford to use those expensive patching compounds, so rough up the surface well with a sledge hammer or a coat of 20 percent muriatic acid. Then install new forms to hold a two-inch topping of fresh concrete. And this time do it right! (Treat it as a new slab, as described in Chapters 1–7.)

One more thing—always wet the area surrounding new or patchwork to prevent too-rapid absorption of water.

Tuck-Pointing

To repoint crumbling mortar, the old mortar must first be removed. Since you don't want to do this too often, you should not only remove the section that is obviously deteriorated, but also any other spots that look crumbly or weak. Large areas of deterioration are most easily removed with a stiff wire brush attachment or an abrasive wheel. Scattered bad spots are just as well excised with a chisel. Start with the bolster for big spots, then work with a cold chisel for smaller areas. A small *cape chisel* may be needed for very small spots.

The crumbled mortar should be cut off to a depth of about an inch, even though the deterioration is only at the surface. The greater depth will give the mortar a place to take hold (this is the same principle as undercutting concrete). Wear safety glasses or goggles when doing any of this type of work to prevent eye injuries.

The mortar mix used for tuck-pointing should be stiffer than the normal mixes described in previous chapters. One part masonry cement is mixed with 2 1/4 to three parts of clean sand, with less water than for your normal use. The mix is about right when it slides from the trowel in a sideways position, but clings when the trowel is turned upside down. Special repointing mortar can be bought at most building supply dealers.

A small plasterer's *hawk* is better for repointing than a regular mortar board, but almost anything that holds the mortar will suffice for small jobs. After dampening all the areas to be repointed, press the mortar into the joint with your trowel. Do the vertical joints first, then the horizontal. Hold the hawk close to the surface and below the work to catch any dripped mortar.

Since the drier mortar that you're using will harden faster than regular mortar, the new joints should be retooled almost as soon as you finish them. Be sure to select a jointing tool that will match the other joints.

The same techniques are used for tuck-pointing block, brick, or stone masonry.

13

Using Decorative Masonry

For building strong, sturdy walls and foundations, regular cored concrete block is an ideal material. But unless you dress it up, it doesn't look awfully pretty. That doesn't mean, though, you should shun concrete masonry entirely. There is, in fact, a wide variety of good-looking and interesting decorative block, as illustrated in this chapter.

The widest use of decorative masonry is for garden walls (Figs. 13-1 and 13-2). This material provides superior privacy, abates noise and strong winds, keeps children and pets in bounds, and generally adds to the value and appearance of your home. You can build it in solid fashion to shut out the outside world completely (Fig. 13-3) or you may use screen block to let some sun and breezes come through.

Although local availability may vary, you should be able to find solid fence designs, screen wall patterns, dimensional wall assemblies, or combinations of solid and screen units for stylish, semi-private uses. Block can be used alone or coordinated with shrubbery, if desired.

The right fence provides dramatic design and a good background for your favorite flowers or vines, too. Most climbing vines cling nicely to concrete.

Fig. 13-1. Attractive screenwalls can be built with decorative block (Courtesy NCMA).

Fig. 13-2. Another example of the attractive screenwall material available with decorative block (Courtesy NCMA).

Fig. 13-3. Even regular cored block can be decorative if used imaginatively. Curved design and stack bond give this solid garden wall real beauty. (Top course should be solid or coped.)

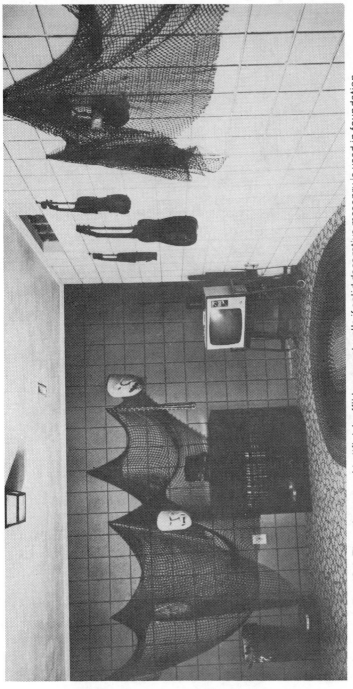

Fig. 13-4. There is no need to "finish off" basement walls if solid decorative masonry is used in foundation construction. These white masonry walls are just as good looking as paneling or other interior finishing material.

Fig. 13-5. Stone-faced concrete masonry makes an attractive and functional fireplace.

Another often-neglected use for decorative masonry is inside the house. It is perfect for basement recreation rooms (Fig. 13-4), or above-ground interior walls. Block has excellent insulating qualities as well as strength. If you build your foundation or walls with solid decorative units, there is no need to build another inside wall. You save money and space.

The many patterns and types are used in different ways, and there is no way to describe all of them. Stone-faced or slush block, for example, has a rustic look that is fine for exterior walls, foundations, and fireplaces (Fig. 13-5). In general, solid decorative units are put up in much the same way as regular cored block. The airier screen types are usually built in much the same manner as brick. Watch out, though, for building codes, and consult your dealer as to how high you can go with the particular type block without pilasters or other types of support. Often, complementary solid units are used for the first and last courses and at regular intervals to provide stability. Your dealer is the best source for information on the particular type you choose.

Only a few of the available units have been shown here, but they will give you some idea as to what can be done. Use your imagination and come up with your own design. You can't go too far wrong.

14

Building A Vinyl-Lined Masonry Swimming Pool

How is it possible to install, in a few days, a swimming pool that is low cost, structurally sound, low in maintenance needs, durable, and beautiful?

The answer is very simple. Build walls of concrete block and install a vinyl liner pretailored to the exact dimensions of the pool size you choose.

Because of the great increase in the number of residential swimming pools that have been installed in the last several years, manufacturers of pool equipment have developed components which are engineered for package type installations. This means that it is no longer necessary to use highly specialized techniques that require the services of specially trained pool contractors.

All that has to be done to construct a top quality swimming pool is to erect concrete-block retaining walls and install the other components which are purchased as a package from your local distributor of swimming pool equipment. There may be a pool contractor in your area who will provide plans and specifications if you buy the liner and other materials from him. The example here was obtained from the Major Pool Co., Clifton, N. J., 07014. With this help, and a reasonable knowledge of laying block, you will become an expert pool builder in a short while.

A concrete block wall swimming pool with a vinyl-lined interior has many advantages, including the exceptionally low maintenance cost and the ease of upkeep. Unlike poured concrete or Gunite pools, the vinyl-lined pool never needs painting. It will not chip or crack due to temperature extremes and its satiny smooth finish will not permit algae to adhere and penetrate. It eliminates water bills because it does not leak and never needs to be refilled.

CHOOSING THE PROPER SIZE SWIMMING POOL

In determining which size and bottom construction to build, a number of factors should be taken into consideration.

Families with small children usually like to allocate as much shallow area for splashing around as is practical. For this reason, the square *hopper-bottom* pool which has half its length in shallow water is preferred. Adults or families with grown children, who are proficient in swimming, may desire to have a greater deep water area available for their activities. The pool which best suits their demends is the *slope-bottom* or the extended hopper-bottom pool. The extended hopper-bottom pool is also ideal for those who wish to do any amount of diving from a board. The back wall of the slope is extended double its normal distance to eliminate the chance of striking the slope when the person is coming up from the dive.

In choosing your pool, never underestimate the bather load. Always select the largest pool that will comfortably fit the space to be allocated. The slight additional cost to construct a larger pool will be many times offset by the greater enjoyment to be had from the additional swimming "elbow-room."

ITEMS TO CONSIDER PRIOR TO EXCAVATION

Preliminary investigation of the local requirements for swimming pools and a clear understanding of construction techniques can save a good deal of time and effort otherwise wasted. A savings in money will also result by avoiding costly errors and changes which may be required later if local regulations have not been followed. Local building codes, regulations, and ordinances vary considerably from one community to another. Check these requirements with the local building inspector and, if necessary, secure the proper permits for construction.

Selecting the Excavation Contractor

Generally, a good local excavating contractor can be recommended by friends or a building official. You will want to have the contractor study the drawings of the particular pool size he is to excavate. Establish the exact requirements of the job and set tolerances to which he is to dig the various sections of the excavation. In this way both you and the excavator will know in advance what performance each of you will be required to make. Make arrangements, at this time, for him to remove any excess soil that will not be needed as backfill.

You may wish to incorporate the performance requirements and pool drawings into a contract so there will be no misunderstandings later on. When engaging the contractor, ask him to assist you in making the excavation layout. He probably has experience in using a transit and can help you with some time- and labor-saving tips.

Location of the Pool

Make a rough layout in the area you think might be the best location for the pool. Stake out the approximate dimensions that the pool will occupy. Then, imagine where additional walks or patio areas will be located. This is the total poolside living space and should be convenient to dressing facilities and house exits. It is desirable to position the pool in the sunniest location possible and to avoid having branches of trees overhead which would tend to shade the pool or litter the water with leaves. If you plan to install a diving board, provide sufficient room behind the deep end of the pool for the board projection and walk area. Avoid placing the pool in a low spot with a high water table because a high water table makes it difficult to form the hopper or deep end of the pool.

Location of the pool filter is important from an aesthetic point of view as well as from the consideration of pump performance. For best efficiency, the length of piping to and from the pool should be kept to a minimum. When making the rough layout of the pool, visualize placement of the filter. Future landscaping should be kept in mind. Most filters can be hidden behind a housing or bush enclosure for compact installations.

Finalizing the Plans

After the pool size and bottom type have been determined, proper location of the water recirculation fittings, skimmer, and underwater light should be marked on the particular pool drawing. It is best to place the skimmer near the center of the long side and downwind of the generally prevailing breezes. In this way the wind will aid in directing surface dirt and dust to the skimmer for removal from the pool. Also, with the skimmer in the middle of the long side of the pool, the vacuum cleaner hose which is inserted into a fitting in the bottom of the skimmer can be kept to a minimum length and both the shallow and deep ends of the pool vacuumed from a central location.

The pool floor construction, whether made of packed sand or concrete, should be determined at this time. If a bottom main drain is to be used, it too should be specified on the plan. A notation should be made to set the drain in four inches of concrete so that it will not shift later and crack the attached plastic piping.

Another important final decision is the type of *coping* that will be supplied and the extent of the concrete deck or patio area. A flush deck can be poured right up to the edge of the pool. If an extended deck is desired around the pool, its shape and dimensions should be determined, laid out in the ground, and then incorporated in a sketch. All of these things should be done prior to the start of excavation so that any alterations can be made at that time. It may be too late to make changes later on.

LAYOUT OF POOL AREA

In order to provide working space when the pool walls are set in place, the outer dimensions of the excavation should be made 18 inches greater, on each side, then the actual pool size. For example, a 20 × 40-foot pool is excavated 23 × 43 feet. Four stakes, one at each corner, are used to outline the excavation (Fig. 14-1). Stakes should be 2 × 2s, good lumber, and at least 3 1/2 feet long. Allowing for the 18-inch working border all around the pool, drive the first layout stake firmly into the ground about 12 inches deep, in the excavation corner which has the highest elevation. From this stake as a working point, lay out the rest of the excavation as described:

Step 1—From stake "A", at the highest ground elevation, lay off the distance of one side of the pool to stake "B". You may wish this side of the pool to be parallel to a house, building line, fence, hedge row, or some other existing element. Care should therefore be taken in setting the direction of this line.

Step 2—With stake "B" as the center point and line 2 as a radius, scratch an arc on the ground at approximate right angles to line 1.

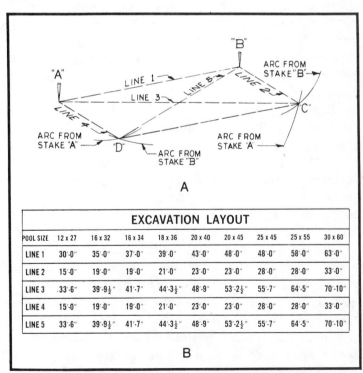

EXCAVATION LAYOUT

POOL SIZE	12 x 27	16 x 32	16 x 34	18 x 36	20 x 40	20 x 45	25 x 45	25 x 55	30 x 60
LINE 1	30'-0"	35'-0"	37'-0"	39'-0"	43'-0"	48'-0"	48'-0"	58'-0"	63'-0"
LINE 2	15'-0"	19'-0"	19'-0"	21'-0"	23'-0"	23'-0"	28'-0"	28'-0"	33'-0"
LINE 3	33'-6"	39'-9½"	41'-7"	44'-3½"	48'-9"	53'-2½"	55'-7"	64'-5"	70'-10"
LINE 4	15'-0"	19'-0"	19'-0"	21'-0"	23'-0"	23'-0"	28'-0"	28'-0"	33'-0"
LINE 5	33'-6"	39'-9½"	41'-7"	44'-3½"	48'-9"	53'-2½"	55'-7"	64'-5"	70'-10"

B

Fig. 14-1. Construction details for pool excavation.

Step 3—With stake "A" as the center and the diagonal line 3 as radius, scratch an arc on the ground to intersect the arc from stake "B." The intersection of these two arcs will determine corner point "C". Drive a stake firmly into the ground at this point.

Step 4—With "A" as center and line 4 as radius, scratch an arc on the ground at right angles to line 1.

Step 5—With "B" as center and line 5 as radius, make an arc on ground to intersect the side arc drawn from "A." The intersection of these two arcs determines corner point "D". Drive a stake firmly into the ground to mark this point.

The excavation layout is now squared. Next choose the level of the top of the pool. This surface will normally be the surrounding walk area at pool side. It is not desirable to have the pool walls protrude more than six inches from the excavation on the low side. If this appears to be the case on a given installation, it may be advisable to dig back some of the earth on the high side for a distance of at least 10 feet from the pool to level the site. This will leave adequate space for drainage. A grade of approximately two inches down for every 10 feet of distance will provide sufficient slope for the runoff of rain and splashout.

Place a reference mark on the stake at the highest ground to mark the position chosen as the top surface of the pool. The next step is to place a corresponding reference mark on the three other stakes to indicate this same elevation. Since no ground is perfectly level, a transit or *dumpy* level will most accurately determine the position of the mark on each individual stake. If no transit or dumpy level is available, a line level may also be used.

If a transit or dumpy level is used, place a *gin* pole—or a 12-foot long, 1 × 2—next to the stake with the pool top reference mark. With the bottom of this pole held even with the stake's reference mark, (*not on the ground*), sight through the transit to the gin pole or 1 × 2 and mark the pole with the level position as seen through the transit. The reference mark on the second stake is determined by holding the gin pole or 1 × 2 next to that stake and raising or lowering the pole until the level mark is even with the horizontal mark in the transit. Then mark the stake where it aligns with the bottom end of the gin pole or 1 × 2. This mark will then be level with the mark on the first stake. This procedure is repeated for all stakes to indicate the top level of the pool.

If a *line level* is used, attach a mason's string at the reference mark on the first stake. Place a level line midway along the string and adjust the position of the string on the second stake until the line level reads level. Mark the stake and tie the string in this position on the stake. Repeat this procedure between each two stakes until all stakes have a corresponding reference mark.

Once the top surface of the pool has been established, the outline of the excavation between the stakes should be made with either powdered lime or flour (Fig. 14-2). This outline will serve as the boundary to which the excavator will dig.

EXCAVATION

A *backhoe* or *Gradall* is the best equipment to use for excavating the pool. This equipment will dig the most accurate hole and disturb the least amount of earth around the hole. It will also make a neat pile of the dirt to be used as backfill and will load a truck to cart away the excess.

Slope-bottom pools are excavated directly to the dimensions shown on the drawing of the particular pool being constructed (Fig. 14-3 and 14-4). Mark the shallow-end and deep-end depth above the level mark on the gin pole or 1 × 2 used to lay out the pool. The excavator is at the proper depth in each section when the corresponding depth mark on the pole lines up with the level sight in the transit as the pole is placed in the excavation. A uniform straight slope from the proper depth in the deep end will assure the correct formation of the pool. The excavation depth may be made one inch deeper than the dimensions shown because too shallow a finished pool will not allow the vinyl liner to stretch out smoothly.

When the specified depths have been arrived at, lay out and square the actual pool size in the excavation according to the horizontal dimensions in the chart. Stakes should be driven in at each corner and strings attached between the stakes.

Fig. 14-2. Before the excavator comes, outline the edges of the pool with lime.

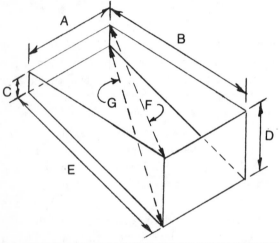

SLOPE BOTTOM POOL LAYOUT						
A	B	C	D	E	F	G
12	27	3'-4"	5'-0"	27'0½"	29'-6½"	29'-7"
16	34	3'-4"	7'-3"	34'-3"	37'-7"	37'-9½"
18	36	3'-4"	7'-6"	36'-3"	40'-3"	40'-5"
20	40	3'-4"	8'-0"	40'-3"	44'-9"	44'-11½"

Fig. 14-3. Typical construction details for slope-bottom pool.

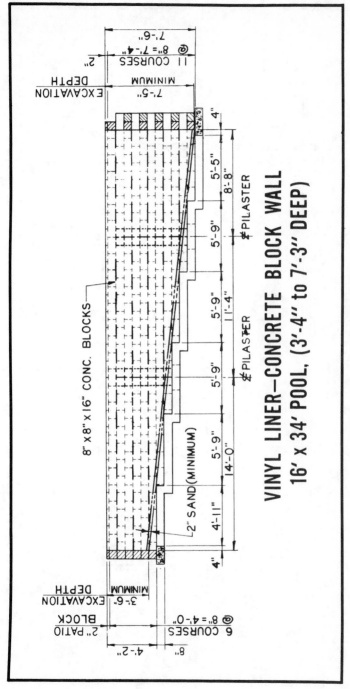

VINYL LINER—CONCRETE BLOCK WALL
16' x 34' POOL, (3'-4" to 7'-3" DEEP)

Fig. 14-4. Further details for slope-bottom pool.

187

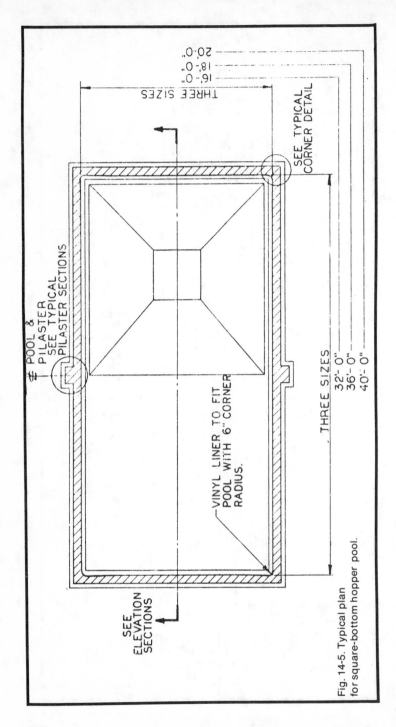

Fig. 14-5. Typical plan for square-bottom hopper pool.

THREE SIZES
20'-0"
18'-0"
16'-0"

SEE TYPICAL CORNER DETAIL

℄ POOL &
PILASTER
SEE TYPICAL
PILASTER SECTIONS

THREE SIZES
32'-0"
36'-0"
40'-0"

VINYL LINER TO FIT POOL WITH 6" CORNER RADIUS.

SEE ELEVATION SECTIONS

Footings should be next laid out, dug, and poured as described for regular block walls.

On hopper bottom pools, the first operation is to dig the entire hole to a uniform depth of 42 to 43 inches below the point selected as the top level of the pool (Figs. 14-5 and 14-6). The depth is indicated on the 1 × 2 which was used to establish the reference mark on the corner stakes. Mark off an additional 42 inches above the level mark on the 1 × 2 and place it in the hole as the excavator digs the various sections. When the 42-inch mark is in the level sight of the transit you know that this part of the hole is at proper depth.

If no transit is used, the depth of the excavation is checked in various sections by making sure that the 42-inch mark on a folding ruler matches up with the "top of the pool" leveling outside the excavation.

When this first part of the excavation has been accomplished, the next step is to lay out the exact pool size on the floor of the excavation leaving the 18-inch border as a working area. This layout should be made very accurately because it will determine the exact final position of the pool. The same procedure for layout and squaring the sides is used here as was used in making the initial square excavation layout. Stakes should be driven in each corner and separate strings attached from one stake to the next to outline a given side.

Step 1—From stake "A", lay off the distance of one side of the pool to stake "B".

Step 2—With stake "B" as the center point of line 2 as a radius, scratch an arc on the ground at approximate right angles to line 1.

Step 3—With stake "A" as the center and the diagonal line 3 as radius, scratch an arc on the ground to intersect the arc from stake "B". The intersection of these two arcs will determine corner point "C". Drive a stake firmly into the ground at this point.

Step 4—With "A" as center and line 4 as radius, scratch an arc on the ground at right angles to line 1.

Step 5—With "B" as center and line 5 as radius, scratch an arc on the ground to intersect the side arc drawn from "A". The intersection of these two arcs determines corner point "D". Drive a stake firmly into the ground to mark this point.

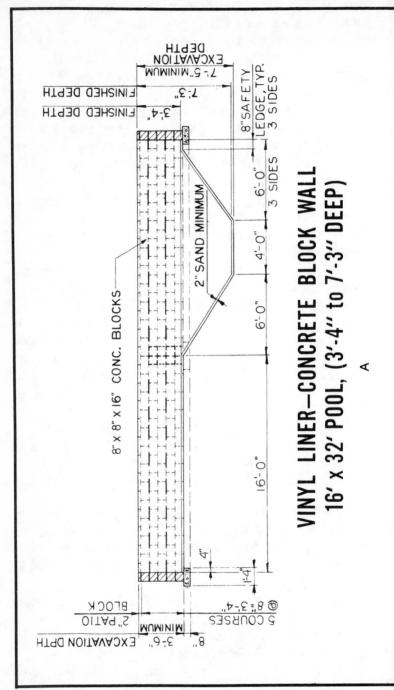

VINYL LINER—CONCRETE BLOCK WALL
16' x 32' POOL, (3'-4" to 7'-3" DEEP)

A

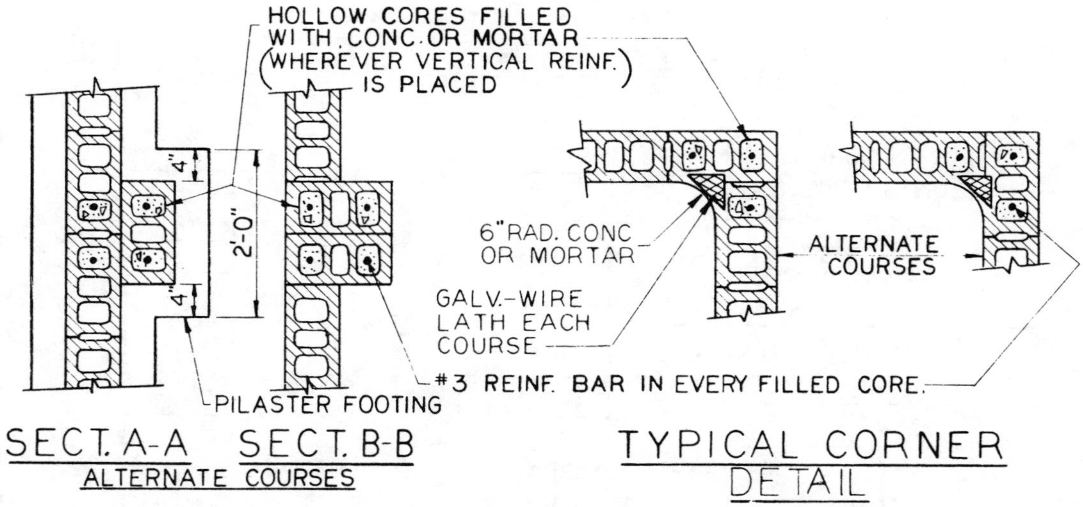

HOLLOW CORES FILLED WITH CONC. OR MORTAR (WHEREVER VERTICAL REINF. IS PLACED)

PILASTER FOOTING

6" RAD. CONC. OR MORTAR

GALV.-WIRE LATH EACH COURSE

#3 REINF. BAR IN EVERY FILLED CORE.

2'-0"

4" 4"

SECT. A-A   SECT. B-B
ALTERNATE COURSES

ALTERNATE COURSES

TYPICAL CORNER
DETAIL

B

Fig. 14-6. (Continued)

191

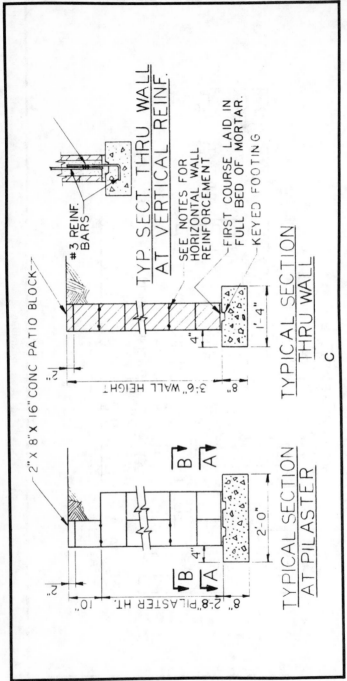

Fig. 14-6. Further details for hopper-bottom pool.

#3 REINF. BARS

TYP SECT THRU WALL AT VERTICAL REINF.

SEE NOTES FOR HORIZONTAL WALL REINFORCEMENT

FIRST COURSE LAID IN FULL BED OF MORTAR

KEYED FOOTING

2" × 8" × 16" CONC PATIO BLOCK

3'-6" WALL HEIGHT

2"

4"

1'-4"

8"

TYPICAL SECTION THRU WALL

C

TYPICAL SECTION AT PILASTER

B

A

B

A

2'-0"

4"

2"

8"

2'-8" PILASTER HT.

10"

Excavation of the Hopper

When the final pool has been laid out with strings at the 42-inch depth, the hopper dimensions should be outlined for the excavator (Fig. 14-7). In order to have the excavator dig the most accurate hopper, it is important to give him guide marks from which to gauge the position of the hopper as he digs. Since the hopper bottom is square—four feet, six feet, or nine feet on a side—its position can be indicated by guide marks on the 18-inch working ledge at the 42-inch depth and also by placing a corresponding set of guide marks at ground level on each side of the excavation opening. Be sure that the hopper layout is eight inches inside the strings of the side and deep-end wall layout to allow for the eight-inch safety ledge. Before the hopper excavation begins, remove the pool layout strings so they will not be covered by dirt from the hopper excavation. Make sure the excavator does not disturb the corner stakes because they will be used again to make the layout of the footings for the concrete block walls.

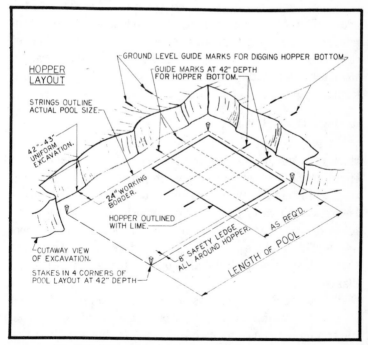

Fig. 14-7. Excavation details for a hopper-bottom pool.

When the excavator digs the hopper he can sight across to the opposite side of the pool and trim the hopper walls to the proper slope as he digs the final depth of the hole. The excavated depth of the hopper should be about two to three inches deeper than the finished pool depth to allow for two to three inches of sand on the bottom as shown on the drawing of the particular sized pool being installed. If you used a transit for the layout of the pool, mark the pool depth, plus two to three inches for sand bottom, above the level position on the gin pole or 1 × 2. Proper excavation depth will have been reached when the indicated mark on the gin pole or 1 × 2 is level with the horizontal mark in the transit sight. If a transit is not used, a simple way to know when the hopper hole is at correct depth is to mark a 1 × 2 with the measurement of the hopper depth below the 42-inch excavation (plus two to three inches for sand bottom), and to run a string across the opening at the 42-inch depth working border. By placing the 1 × 2 in the hopper hole as the excavator digs, the correct depth will be indicated when the mark on the 1 × 2 matches up with the level of the string across the opening. If frequent measurements are taken of the depth and slope of the hopper walls, this part of the excavation can be controlled accurately. A smooth liner installation is almost entirely dependent upon accurately positioning and sizing this hole.

Footings for Block Walls

On sloping bottom pools, it is necessary to provide footings which step down eight inches at varying distance along the length of the pool as shown on each particular pool drawing. After laying out the exact pool size in the excavation, stake out the trench which is to be dug for pouring the footings. The inside of this trench should be four inches inside the pool layout. The width of the trench should be 16 inches except where pilasters are located as shown on the pool drawing and pilaster details.

If an accurate trench is dug for the footings, it is not necessary to frame the inside and outside edges of the footings. The trench will serve as its own frame. In this way there are no voids to fill between the footing and pool floor. Where the

194

footing steps down, however, it is necessary to provide framing across the step width and this framing should be made accurately both in elevation and in longitudinal distance from the ends of the pool. Accurate placement of the pilaster footings is also very important to the structural support of the pool walls.

Footings for walls of hopper-bottom residential pools are poured level all around the pool and the accurately dug trench can serve as its own framing. Be careful to trench out properly where the pilaster footings are indicated.

As discussed in previous chapters, concrete footings should have a width equal to twice the wall thickness and a depth equal to the wall thickness. While the concrete is still soft, roughen the top of the footing surface and scratch out a keyway approximately one-inch deep by four inches wide in the center of the footing to promote a better bond at the joint between the footing and the wall. Also, reinforcing bars for the vertical reinforcement of the block wall should be placed near the center of the footing width to insure maximum efficiency against water pressure from the pool side of the wall and against the earth pressures from the back side of the wall. It is suggested that a small longitudinal rod be provided along the re-bar line near the top of the footing so that dowels can be accurately spaced and securely tied in the correct position. Re-bars should be placed at least every four feet along the unsupported length of the wall and in each corner and pilaster as shown on the drawing details.

WALL ERECTION

After the footing has been continuously poured around the bottom of the pool excavation, start erection of the pool walls. The first row of concrete blocks should be laid in a full bed of mortar so that the walls are completely anchored to the footings. This will prevent any shifting of the wall due to water or earth pressure after the pool is filled. The reinforcing bar placed in the footing should extend about 18 inches above the footing. This will leave enough protruding so that the first row of block will be held by this rod and another, longer rod, can be placed to the first one with 16-inch wire. Knock out ends of block units as required to fit around vertical bars in place. As additional blocks of the wall are laid and the cores through which the rods pass are filled with mortar, the wall will be securely tied together as a continuously reinforced vertical unit.

The first row of block on the footing should be positioned very accurately and all dimensions checked and double-checked. Be sure the pool is square. (See previous chapters.) Each block should be carefully leveled and plumbed. Then continue laying the concrete block, one course on top of another, with each vertical joint over the center of the block below.

As the walls go up, keep making frequent checks on all dimensions and be sure that the walls are straight and plumb. Install the proper wall fittings for water recirculation, lights, and skimmers in their appropriate locations, as the wall is being erected. Do not wait to cut these openings in after the wall is finished because they are much more difficult to install at that time. You should have determined the placement of the skimmer and water recirculation fittings by marking them on your pool layout drawing before wall construction was started.

Horizontal wall reinforcement should also be added as the wall goes up. Place one 1/2-inch diameter steel rod in a full, continuous bed of strong mortar one inch in from the front edge of the block. Place another rod one inch from the back edge of the block. The rods should be made continuous by overlapping and splicing 18 inches of two joining rods with 16-gauge wire. Both the front and back rods should be bent and carried continuously around corners. Prefabricated joint reinforcement with nine-gauge side wires may also be used with corners bent and cut according to manufacturer's recommendations.

The horizontal wall reinforcement should be placed in every other course starting with the first course above the bed course. On pools with a sloping bottom, where step footings are employed, the first joint with the horizontal reinforcing rods should be the first course above the bed course in the lowest part of the pool. It is not necessary to place horizontal reinforcing rods in the joint under the top course of two-inch concrete patio blocks on pools where this joint would normally receive the rod.

Vertical reinforcing rods should be placed in the hollow cores of the blocks that form pilasters and corners (as shown in the details on each pool drawing) as well as at four-foot intervals along unsupported sections of the walls. Wherever vertical reinforcing rods are placed, a 3/8-inch diameter (#3) rod should be used and the core filled completely from top to bottom with concrete or mortar. The method of securing the rod in the footing and splicing additional rod above the footing is shown in detail on each pool drawing. Vertical rods should not protrude above the last course of 8 × 8 × 16 block because the top two-inch patio block is a solid block. Cores that are to be filled should, therefore, be filled before this top course of block is laid.

When the complete wall has been erected, it is absolutely essential that all excess mortar be removed from between the joints and wherever it has adhered along the inside wall. The joints should then be troweled smooth. After this has been done, coat the inside of the wall with a 1/8-inch slush coat of cement plaster. This coat evens out any irregularities or imperfections in the blocks or wall and leaves a smooth surface with no rough edges which may abrade or damage the vinyl plastic liner. The plaster consists of one part of portland cement and two and one-half parts of very fine mortar sand with the addition of enough water to produce a smooth, flowing mixture. Wall surfaces should be free from dirt, oil, dust, or any material that might prevent satisfactory bond. Wall surfaces should be moistened prior to application of the slush coat, which should be troweled smooth and allowed to harden for 24 hours before liner installation. On pools with concrete bottoms, the junction of the wall face and the pool bottom should be thickened to form a two-inch radius.

Once the walls are up, finish grading the pool bottom to the exact size. If the excavator has worked carefully, and the footings and walls are erected at their proper depth, the pool floor should also be very near its exact depth. Check this dimension by running a string line across the top of the pool wall at various points and measuring the depth. A small amount of hand trimming is all that should be required, at most, to bring the bottom to proper depth.

As mentioned, this depth may be made one inch deeper than required but may *not* be made less than the dimensions shown. This final sizing of the pool bottom governs the smooth future appearance of the liner in the finished pool.

POOL BOTTOM PREPARATION

Slope Bottoms

The next step involves preparation of the pool's bottom. For slope-bottom pools, measure down from the top of the wall in the shallow end and mark the corners with a chalk mark at the finished pool depth. Measure down from the top of the deep end wall and mark each corner with a chalk mark at the finished deep-end and deep-end walls to mark the entire wall with this dimension. Then snap a chalk line to make the proper slope of the length of the pool from the shallow to the deep end on each long wall. These chalk lines will serve as a guide for the level to which sand can be tamped and rolled to form the pool floor to the correct depth. This depth can be one inch deeper than shown on the drawings but cannot be shallower.

Laying about two to three inches of a thoroughly washed, dead sand on the bottom of the pool will provide this depth. Wet and tamp solid all loose earth in the pool bottom before adding the sand. Dump the sand near the center of the shallow-end wall, and rake it down the slope. Particular care should be taken with the tamping of the sand bottom. The smoothness and evenness of this bottom will determine the appearance of the liner when it is placed in the pool. In order to get maximum compaction and evenness of surface, it is suggested that the sand be drenched thoroughly and deeply with water before and after placement in the pool. This will allow the finishing operation to proceed in orderly fashion, and will prevent one section of the bottom from drying out before the whole is completed and the liner set in place.

To attain a thoroughly compact sand bottom, the wet sand should first be tamped, then rolled with a garden roller. Two men can do this job. Compaction work should progress from the deep end toward the shallow end. Create a two-inch *fillet* of sand at the base of the walls all around the pool. This is done by placing sand at the wall base and shaping the fillet to its proper radius with the side of a 4-inch diameter milk bottle. This fillet will give a smoother finish to the liner at the base of the wall than would a sharp base angle and will also take stress off the liner at this point. When the workers are finished tamping at the shallow end, they should tamp smooth their last tracks as they leave the pool.

One more thing should be done to insure the best appearance of the pool: For the last finishing operation on the sand bottom, attach a long rope to each side of a piece of one by three-foot carpeting, and drag the carpet across the entire bottom by having one person walk along the outside of each wall and pull the carpet along. This procedure will even out any slight imperfections on the sand surface. If no carpeting is available, a wet towel may be used instead.

Hopper Bottoms

On hopper-bottom pools, follow the same procedure of snapping a chalk line at the finished pool depth on each wall all around the pool. This line will be at the same level on each wall. Then, refer to the drawing of the particular pool being installed, and locate the position of the square hopper bottom (Fig. 14-8). Fabricate a square jig of 1 × 2 lumber either four feet, six feet, or nine feet on each side, according to the hopper bottom of the particular pool being installed. In order to set this jig in the exact hopper bottom, attach two strings across the hopper as shown. The strings will intersect in mid-air over the hopper. By dropping a plumb line from the intersection, the exact corner of the hopper will be indicated. Place the corner of the 1 × 2 jig in position under the plumb line. Then move the plumb line along the string the distance "S". This plumb will accurately place the second corner and the bottom jig will be in perfect relation to the pool sides. The top surface of this jig should be set at the finished surface of the sand bottom. Use shims with the jig to establish this height accurately.

The next step is too set guide boards in each corner of the hopper side walls. These four boards should be 1 × 2s and their length is determined according to the chart for each pool

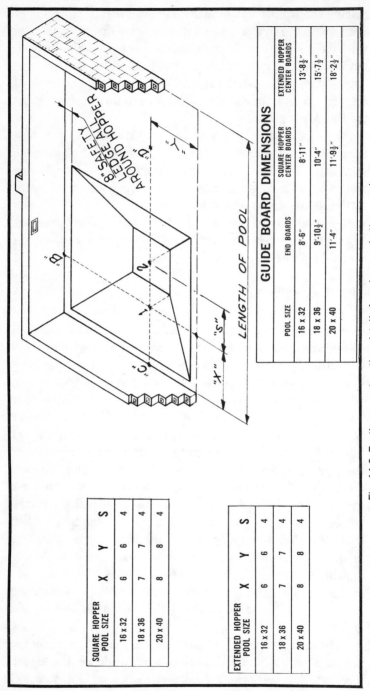

GUIDE BOARD DIMENSIONS

POOL SIZE	END BOARDS	SQUARE HOPPER CENTER BOARDS	EXTENDED HOPPER CENTER BOARDS
16 x 32	8'-6"	8'-11"	13'-8½"
18 x 36	9'-10½"	10'-4"	15'-7½"
20 x 40	11'-4"	11'-9½"	18'-2½"

SQUARE HOPPER POOL SIZE	X	Y	S
16 x 32	6	6	4
18 x 36	7	7	4
20 x 40	8	8	4

EXTENDED HOPPER POOL SIZE	X	Y	S
16 x 32	6	6	4
18 x 36	7	7	4
20 x 40	8	8	4

Fig. 14-8. Further construction details for a hopper-bottom pool.

200

size. The bottom of each guide board should be positioned flush with the corner of the hopper bottom jig and the top of each board should be at the top of the hopper's finished sand depth. The top of each board should also be eight inches away from the pool wall to leave room for an eight-inch safety ledge. With the top surfaces of the guide boards and jigs as references, finish the hopper by final hand trimming of the undisturbed earth.

As final grading progresses, wet down and tamp solid all loose earth in the pool bottom before adding sand. To save effort when adding sand, dump it near the center of either shallow-end sidewall. A certain amount will roll down. The balance will be easy to rake down. Particular care should be taken with the tamping of the sand bottom. The smoothness and evenness of this bottom will determine the appearance of the liner when it is placed in the pool. In order to get maximum compaction and evenness of surface, a thoroughly washed, dead sand should be drenched completely with water before and after placement in the pool. This will allow the finishing operation to proceed in orderly fashion, and will prevent one section of the bottom from drying out before the whole is completed and the liner set in place.

To attain a thoroughly compact sand bottom, the wet sand should first be tamped then rolled with a garden roller. By attaching ropes to the roller, the sloping hopper sides can be tightly compacted by guiding the roller up and down the slopes from an outside position at the top of the wall sections. Two men can do this work.

Sand depth should be two to three inches. After the hopper side walls have been tamped, remove the corner guide boards and bottom jig. Fill the voids with wet sand, retamp and reroll the whole hopper to final smoothness. Be sure to provide the eight-inch safety ledge around the hopper.

The man or men working on the sand bottom of the hopper and tamping the sloping hopper sides must be provided with a means of getting into and out of the pool without disturbing the compacted areas already worked on. For this purpose, a plank walkway, with *furring strip treads*, should be laid up on the slope of the hopper side. The workmen may then walk up the incline without leaving footprints. When the tamping operation has been completed, the workmen leave a garden roller, with guide ropes attached, in the bottom of the hopper. They then leave the hopper by means of the plank walkway and pull the plank out after them. The garden roller can then be pulled up the hopper incline over the place where the walkway was used, it will smooth out any depression which may have been left there. Finish the shallow end of the pool by rolling and tamping the sand to the proper depth and uniform smoothness.

At the base of the wall, all around the pool, create a two-inch *fillet* of sand. This is done by placing sand at the wall base and shaping the fillet to its proper radius with the side of a four-inch diameter milk bottle. This fillet will give a smoother finish to the liner at the base of the wall than would a sharp base angle. It will also take stress off the liner at this point.

Now the pool bottom and hopper should be at the exact dimensions and shape that the final pool will have. One more thing should be done to insure the best appearance of the pool: for the last finishing operation on the sand bottom, attach a rope at each side of a piece of one by three-foot carpeting. Drag the carpet across the sand from outside the pool. This procedure will even out any slight imperfections on the sand surface. If no carpeting is available, a wet towel may be used instead.

INSTALLING THE LINER EXTRUSION

While the sand bottom is being rolled and tamped, the rigid plastic extrusion, into which the liner fits, may be installed on the top of the pool wall with concrete nails. The four radius corner pieces should be installed first. Fasten these to the top two-inch patio block by driving concrete nails through the plastic into the block near each end, in the center tab at the back and at each side of the radius. The nails need not be driven into the block itself but can easily be hammered into mortar between adjoining blocks while it is still green, and into the concrete mortar with which the corner radius is formed.

Make sure that the inside edge of the plastic extrusion is flush with the inside edge of the pool wall. Next install the long straight lengths of extrusion into the mortar between the joints all around the pool. This will place a nail at 16-inch intervals. Butt the lengths of extrusion together tightly. Excess from the last piece can be trimmed with a hacksaw after it is installed.

The extrusion is further anchored into position by installing the coping at this time before the liner itself is installed. If a precast stone coping is used, the extrusion will be completely anchored in mortar. If a deck is to be poured flush with the inside surface of the pool, an inch of mortar may be placed on top of the patio block also covering the extrusion. In this way the extrusion will be poured as one unit later on.

INSTALLING THE LINER

Before installing the liner, double-check the sand bottom to be sure there are no stones, twigs, or sharp objects that might damage the liner. Remove any such objects and tamp sand into the holes.

Installation of a plastic liner is always easier when the weather is warm and the sun is shining. On such days, the liner can be handled better and the packing wrinkles will disappear more readily. If the liner is to be installed in cool weather, store it in a heated room or basement for a day or two prior to installation. This will allow the liner to retain some warmth as it is unrolled at the pool. The wrinkles will not be so sharply pronounced as when the liner is unrolled cold.

Four or five people are needed to install the liner. The liner is shipped in a large drum. Roll the drum next to the pool and remove the liner carefully, being sure that it does not hit any sharply protruding edges on the package or on the ground. Unroll it lengthwise toward the deep end of the pool and be sure there are no sharp objects on the ground that may abrade the liner. Remove the liner from the drum two or three hours before the time planned for installation. This gives the vinyl a chance to warm up so that most of the packing creases will disappear and make the liner easier to handle and install. Make sure the edges and corners of the liner are accessible.

Place two people at each end of the unrolled liner. One person at each end should take hold of the liner corner that will be positioned at the opposite side of the pool; the other person should hold the liner corner that is in its correct place. The two people holding the opposite-side corners then walk across their respective ends of the pool unfolding the liner like an accordian. The liner will then be suspended over the excavation with all sections in their proper relative positions.

All fitting and placement is done from the outside of the pool. Otherwise footprints would be created in the sand bottom which would be impossible to erase after the liner is installed.

Take the top edge of the liner, and starting from the corners, insert the bead into its mating plastic extrusion all around the pool (Fig. 14-9). The exact corners are marked with ink on the top underside of the liner so they can be positioned in the middle of the radius of the corner.

Adjust the liner and pull it gently into position so that the bottom fits exactly in the excavation. On slope-bottom pools, the seam between the walls and the bottom should be positioned in its proper location all along the entire slope of the pool. The bottom should be flat without any excessive folds in any section.

Hopper-bottom pools should have the hopper seams matched up with the corners of the sloping hopper walls in the deep end of the pool. Keep pulling on the liner; always grab a heavy fold of material with both hands and pull in short tugs. Don't be afraid to pull the liner into position. The heavy gauge vinyl will take a lot of pull without damage.

When you are satisfied that the liner is in proper position, remove six inches of the liner head bead from the extrusion in one corner of the deep end. In this opening, insert a vacuum cleaner hose to about 18 inches below the top of the wall. It is best to use an industrial-type, heavy duty vacuum rated for continuous duty. A household vacuum can also be used with its filter bag removed if no industrial type is available. Stuff wet rags into the opening where the hose is inserted between the liner and wall, so that air does not enter and decrease the vacuum. Also, the openings for wall fittings may be sealed off with cardboard and masking tape to increase efficiency of the vacuum. Back fill behind the wall with moist earth to a depth of about six inches and tamp thoroughly. This will provide a seal against air leakage from underneath the wall.

Fig. 14-9. Liner is fit into plastic extrusions.

Start the vacuum cleaner. In about 15-20 minutes, the liner will be drawn tight against the pool excavation and side walls. You will immediately see if the liner bottom is seated exactly in the excavation bottom or if it needs repositioning. Make adjustments where necessary (if possible while the vacuum cleaner remains running) by pulling the liner into place. If folds or wrinkles are present, they can easily be eliminated at this time by smoothing them out with a pool brush. In some cases, where substantial repositioning is required, it may be advisable to shut off the vacuum cleaner and make the adjustments before drawing the liner into position again with the vacuum cleaner.

After the liner has been properly positioned, start filling the pool with water while the vacuum cleaner is still running. On slope-bottom pools, continue adding water and brush out any wrinkles which may form as water rises up the slope. The side walls will seldom, if ever, show any wrinkles. On hopper-bottom pools, add water until the hopper is full and the shallow end is covered by one or two inches of water. Brush out any wrinkles on the shallow bottom during this filling operation when about one inch covers the bottom. When the wrinkles are removed, shut off the vacuum cleaner and continue filling with water.

INSTALLING THE SKIMMER AND LIGHT

Each wall fitting is ordinarily supplied with two gaskets. Before the liner is put in place, one gasket should be cemented to the fitting. When the liner is installed and the water level reaches two to three inches below the recirculation fittings, skimmer, or light, these elements may be installed. By pressing on the liner at the various openings, locate the screw holes and puncture them with a sharp nail or ice pick. Install each of the various components, making sure that one of the gaskets supplied with each of the fittings is placed between the fitting and the liner, and the other gasket is between the faceplate and the liner, as shown on the fitting details. The liner obstructing the opening should be cut out with a very sharp knife after the faceplate is installed.

BACKFILLING

Begin backfilling slope-bottom pools immediately and hopper-bottom pools when the water level reaches the walls. Backfilling should be done at approximately the rate of the water rise. Do not allow the height of the backfill to exceed the height of the water by more than 24 inches. Do not allow the height of the water to exceed the height of the tamped backfill by more than 12 inches. Backfill should contain a minimum of clay and absolutely no shale or large rocks. As the backfill is put behind the wall sections, it should be tamped into place by walking on it or by using a tamper. It may be wet or wetted down while being placed in the excavation. Wetting will hasten the settlement of this area if a patio is to be constructed soon after pool installation. If some sand is left over from the bottom construction it may be used as backfill. It will compact very well. The bottom layer of backfill should be porous material to allow proper drainage away from the back side of the wall.

INSTALLING PLASTIC PIPE

Use a good grade of #75 NSF-approved plastic pipe, recommended for freezing conditions. As the backfill reaches the fittings, connect the plastic pipe to the skimmer suction fitting and to the recirculation fitting. To prevent a possible air leak, the suction line from the pool to the pump should slope uniformly in one direction without any kinks or dips. Also maintain uniform slope on the skimmer line.

To install plastic pipe, place plastic joint compound on the threads of the adaptor and screw it into the fitting. Then put some compound on the other end of the adaptor, slip two pipe clamps over the end of the plastic pipe, and push the plastic pipe over the serrated end of the adaptor until it is well in place. Slip the clamps over the adaptor and tighten them 180° apart. These clamps will take a *set* in about an hour and should be tightened again at that time. When handling the pipe, avoid overbending or kinking it. If it is to turn a corner, make a large radius bend. If sharp turns are necessary, use plastic elbows.

When installing recirculation piping, it is very desirable to support this piping completely. This is easily done if the plastic pipe is laid directly on the undisturbed earth of a trench dug between the pool and the filter. However, where the pipe must cross the gap from the outside of the excavation to the pool wall, some additional support should be provided. Otherwise, when the backfill settles, a void will be created under the plastic pipe. The pipe alone must then take the strain of the soil pressure above it. Pipe failure may result because plastic pipe is not manufactured to endure this stress. To eliminate the possibility of such a failure, it is recommended that concrete blocks be placed under the pipe as supports between the excavation and the pool wall fitting.

PRECAST OR STONE COPING

Patio block or regular precast stone swimming pool coping may be attached to the top of the pool by applying mortar or grout and setting the coping. Care should be taken not to use excessive grout which will spill over into the pool when the coping is pressed down. You will need corner coping with a six-inch radius.

FLUSH DECK AT POOLSIDE

An alternate to placing stone coping around the edge of the pool is to pour a concrete deck right up to the edge of the pool. This is done by providing a wood form to hold the concrete from spilling over into the pool. The inside of the form is positioned just over the edge of the plastic extrusion and secured in the ground behind the pool so that the finished deck will be even with the inside edge of the pool. Remember to form a six-inch radius in the corners so the deck will conform to the pool at these points. It is advisable to pour an expansion joint about 12 inches behind the pool edge so that the main portion of the deck can move independently of the pool walls when freezing and thawing of the earth cause this motion. For the same reason, a separate deck pad should also be poured around the top of the skimmer and deck box for the light. These fittings should not be controlled by the pressures exerted by the earth on the deck.

INSTALLING WALKS

It is best not to lay permanent concrete walks immediately on fresh backfill. If convenient, first lay a walk of patio block set in level sand around the perimeter of the pool. After the backfill has settled, the blocks may be lifted and set in more permanent foundation, or concrete walks may be poured. When laying the walks, be sure that provisions are made for ladder or diving board anchors if they are to be used with the pool.

CARING FOR VINYL–LINED POOLS

Vinyl liner material is specially formulated for swimming pool use and should last many years. Dirt that collects at the waterline from hair oil and suntan lotions is easily removed with a mild household cleaner (not containing chlorine), or with a detergent. Sparkling water clarity is simple to maintain through the proper use of the filter system in combination with adequate chlorination. Your local swimming pool equipment distributor has complete information on the proper treatment of swimming pool water and will be glad to assist you.

Continuing care of your pool also includes adequate protection and treatment during the non-swimming season as well as during the summer months. Before cold weather sets in, all components of the pool should be winterized.

The vinyl liner and concrete block walls also need to be considered in the winterizing operation. When freezing temperatures cause the water and moisture in the ground behind the pool to solidify, the earth in this section expands and pushes inward against the concrete block walls. Therefore, it is essential that water be left in the pool during the winter to freeze and expand. This outward force of the pool ice on the wall will offset the inward force of the surrounding ground. The balance between the inward and outward pushing forces on the wall keeps the wall from being damaged. It is not necessary to place logs in the pool. They only make the pool dirty.

A completely full and frozen skimmer, however, might tend to crack because of the expanding forces against the cast aluminum housing. Therefore, insert a block of non-absorbent, yet compressible plastic (such as polyurethlane foam) into the skimmer body and it will be protected against failure due to ice expansion. Keeping the pool full will also extend the useful life of the liner because the water and ice will protect it from being damaged by falling objects and careless handling.

DO NOT DRAIN THE POOL

Reseating a liner in a pool from which the water has been drained is always a difficult operation. The holes in the liner for the wall fittings do not normally all line up exactly in their former location and the tamped sand bottom must be remolded before the liner can be reinstalled. All in all, draining a vinyl-lined pool would create difficulties and is unnecessary.

It is extremely desirable to keep all foreign matter out of the pool during winter and to maintain the water in its disinfected state during this time. A protective plastic skin over the entire pool surface is designed to eliminate contamination from dirt, sand, soot, small animals, and insects. The skin will also completely discourage the possibility of algae being introduced into the water during the non-swimming months.

The key to such absolute protection is total contact between the plastic and the deck area around the pool. A good cover has four separate, wide water sleeves to distribute as much as one ton of water at the perimeter of the pool and form a continuous, positive seal between the cover and deck. Each side should have a separate water sleeve with its own filling valve conveniently located at the cover corner.

In order to derive maximum benefit from your cover and to maintain water clarity throughout the fall, winter, and spring, it is recommended that you superchlorinate to 5 to 10 times the usual dosage of chlorine just prior to installing the cover. This large residual will destroy all organic matter in the water so that the pool is completely disinfected. Make sure that the major portion of the cover rests directly on the pool water surface. In this way the chlorine will remain effectively in solution.

In the spring, when occasional warm weather could promote the growth of algae, lift up one corner of the cover and check the chlorine residual with your test kit. If the chlorine has been consumed over the winter, rechlorinate to about two times the usual dosage. Replace the corner of the cover in position and your water will remain crystal clear until you are ready to open your pool up for the new swimming season.

Part 3 : Brick

15

Modern Brick—Tools and Materials

After stone, brick is man's oldest building material. It was the first *manufactured* building material, and evidence exists that it was used by early man at least 5000 years ago. There are also indications that a brick *guild* existed when King Solomon's Temple was built about 1000 B. C.

Early types of brick were usually *adobe*, a type of brick which is still used today, particularly in the American Southwest, Egypt, and Asia. Adobe is easy to make. Clay is simply put into a mold and dried in the sun. But *burned and fired* bricks were used long ago, too. These were harder and stronger. Many were glazed so masterfully that some bricks recovered from excavations retain their brilliant red, yellow, or blue-green colors.

The brick of yesteryear was considerably larger than modern brick. In the ancient and fabulous city of Babylon, the burned bricks were twelve inches square and three inches thick. Adobe brick sizes varied from six to sixteen inches square and two to seventeen inches in thickness. Like the burned brick, most early adobe brick was also square.

Like the craft unions of today, the early guilds guarded their art jealously. The bricklayers made their own brick, and the technique was a trade secret, handed down to sons and trusted apprentices. Somehow, during the handing-down process, some of the secrets were lost to posterity. The formulas for glazing, although brought to a fine art before the birth of Christ, were submerged somewhere and weren't rediscovered until the past century.

Essentially, brickmaking hasn't changed since those early days. Clay or shale is mined (the art is called *winning* in the industry), pulverized and tempered with water (the acts of *grinding and pugging*), formed, dried, and burned. For many years, these processes were all done by hand, but machinery has taken over most of these steps during the past two centuries. A brickmaking machine was invented in 1800, and now the entire process is mechanized (Fig. 15-1).

In modern bricklaying plants, the tempered clay is extruded, much as toothpaste is forced from a tube, cut into the desired lengths, and dumped into molds. The molds are previously coated with sand or water to prevent the clay from sticking. After forming, the brick is placed in kilns for drying and baking, also called *burning*. (Brick, at one time, was literally burned by building a fire under stacked brick.)

Some extruding machines contain dies for making holes or cores in the brick, to reduce weight. Coring should not exceed 25 percent. Glazed brick is made either by spraying on glazes and then burning them at high temperatures, or by applying the glaze after burning, then refiring. Almost any color is possible by using the latter method. Glazed brick has a ceramic, glass-like coating which is hard and impervious to moisture.

MODERN BRICK

Modern brick comes in many sizes and shapes, and each has its own special uses (Fig. 15-2). Many of the longer bricks, such as Roman and Norman, are designed for architectural effect rather than for any special utilitarian purpose. Others, such as engineer's double and triple brick, are made more for strength and other specific purposes.

The do-it-yourselfer is well advised to stay with modular *standard* or *common* brick, which is nominally four inches thick, eight inches long, and 2 2/3 inches high. (See Table 15-1).

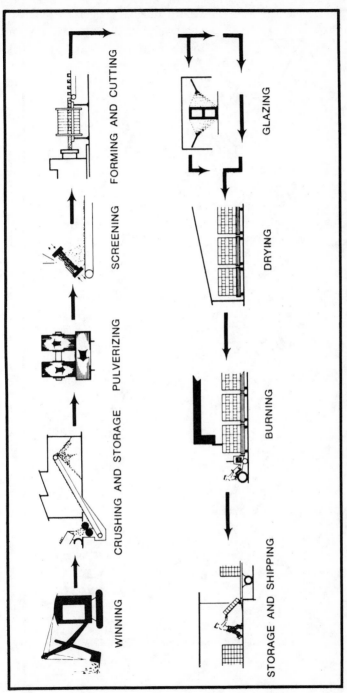

WINNING CRUSHING AND STORAGE PULVERIZING SCREENING FORMING AND CUTTING

GLAZING DRYING BURNING STORAGE AND SHIPPING

Fig. 15-1. How modern brick is manufactured.

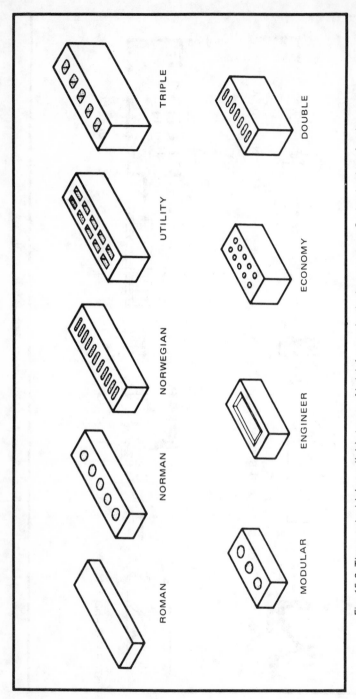

Fig. 15-2. The most widely available types of brick for use by the homeowner. Coring is given for the example only, and will vary with types according to the individual manufacturer. "Common" or "standard" brick may or may not be modular.

ROMAN   NORMAN   NORWEGIAN   UTILITY   TRIPLE

MODULAR   ENGINEER   ECONOMY   DOUBLE

As with concrete block, the nominal dimensions allow for mortar, so that the actual dimensions are 3/8 to 1/2 inch less, depending on the desired size of the mortar joint. Unfortunately the brick industry, though it is attempting to standardize its measurements, is not completely in agreement on the optimum size for *standard* brick. You may find that local standard brick is not modular, but exactly 3 3/4 × 8 × 2 1/4 inches. When buying brick, measure it yourself to make sure that it will fit into your plans. If you can, buy *modular* brick—that is, nominally 4 × 8 on the big side, counting mortar, so that two ends or half-bricks will equal the long side of a whole one. The reason for this is the same as for planning concrete block in modular units—to avoid unnecessary cutting and fitting.

The height of the brick is of less importance than the length and thickness. Much standard brick is made in exact 2 1/4-inch heights instead of 2 2/3 inch modules. This is usually acceptable, since this dimension plays little part in planning, but keep it in mind when counting courses.

Be particularly careful when buying brick for use without mortar, such as for a brick-in-sand patio. Many publications give plans for this usage without considering the fact that no mortar is used. If you buy common non-modular brick, often the only kind available, you will find that herringbone or other attractive patterns are very difficult to lay because the brick is not exactly twice as long as it is wide. This can cause a lot of extra work aligning the brick to compensate for the extra length. For horizontal, non-mortar jobs, use either running bond or special patio brick, which *is* exactly twice as long as it is wide. (For more on this, see the next chapter.)

MORTAR

Since bricks are considerably smaller than concrete blocks, there is proportionately a much greater volume of mortar in the average installation. Mixing and applying mortar is therefore of even greater importance than with concrete block. The professional bricklayer uses several different types of mortar. (See Table 15-2). The handyman, however, is almost always safe in using a standard premixed

Table 15-1. The Different Types and Sizes of Bricks

Unit Designation	Dimensions			Modular Coursing
	Thickness	Height	Length	
Standard or Common *	4"	2-2/3"	8"	3C = 8"
Engineer	4"	3-1/5"	8"	5C = 16"
Economy	4"	4"	8"	1C = 4"
Double	4"	5-1/3"	8"	3C = 16"
Roman	4"	2"	12"	2C = 4"
Norman	4"	3-1/5"	12"	3C = 8"
Norwegian	4"	3-1/5"	12"	5C = 16"
Utility	4"	4"	12"	1C = 4"
Triple	4"	5-1/3"	12"	3C = 16"
SCR brick * *	6"	2-2/3"	12"	3C = 8"
6" Norwegian	6"	3-1/5"	12"	5C = 16"
6" Jumbo	6"	4"	12"	1C = 4"
8" Jumbo	8"	4"	12"	1C = 4"

* Usually more widely available in non-modular sizes, exactly 3-3/4" × 8" × 2-1/4".

**A patented slotted brick type used mainly by professional bricklayers.

218

Table 15-2. The Different Types of Mortar

Type S: general-use mortar, recommended for do-it-yourself applications, and particularly where high lateral strength is a factor. Consists of one part portland cement, one-half part hydrated lime, and four parts clean sand (by volume).

Type M: high-strength mortar, also for general use, but specifically recommended for masonry below grade and in contact with earth, or where maximum compressive strength is required. Consists of one part portland cement, one-fourth part hydrated lime, three and one-half parts clean sand.

Type N: medium-strength mortar, for general use above grade where no special strength is required. Consists of one part portland cement, one part lime, five parts clean sand.

Type O: low-strength mortar, suitable only for use in non-load-bearing walls of solid units, or interior non-loadbearing walls of cored brick. Consists of one part portland cement, two parts lime, seven and one-half parts clean sand.

All mortar, of course, is mixed with water as described in the text. Sand proportions may vary somewhat, but should never be **less** than 2 1/4 parts or **more** than three times the sum of the volumes of cement and lime.

masonry cement, or in mixing his own in proportions of one part portland cement, one-half part lime, and four parts clean sand. Water is added to the dry mix until it reaches a mud-like consistency.

Brick mortar is quite similar to the mortar used for concrete block, and has a useful life of about 2 1/2 hours. No more mortar should be mixed than you can use in that time, and it should occasionally be retempered by adding enough water to compensate for evaporation.

A common fault with mixing mortar is oversanding. This is due to the fact that shovels are used to scoop up and add the sand, whereas the cement and lime come in quantities that are more accurately defined. Since the capacity of shovels varies widely, find out the size of an average shovelful by constructing a box one foot in all directions. Experiment to find out how much shoveling it takes to fill this cubic foot.

TIES

The metal ties used in brickwork are the same as, or similar to, the ties described for concrete block. Corrugated metal used for tying brick veneer to wood-frame construction are made of 22-gauge galvanized steel and should not be less than 7/8" wide. Dovetail ties used with concrete or block should be 16 gauge and 7/8 inches wide. Z-bars and rectangular ties are also used for concrete masonry.

Always consult building codes before planning brick veneer or double-tiered walls. The size of the ties can make an important difference. Many codes specify that the minimum thickness of any masonry wall shall be eight inches. A brick wall attached with thin ties cannot be calculated as part of the eight inches, and is considered a veneer. If heavier ties are used, however, the brickwork can be considered part of the actual thickness of the wall and added to its dimension. If, for example, the veneer is four inches thick, and the requirement is for an eight-inch wall, the backup must be eight inches by itself when thin ties are used. If heavier ties are used, however, the backup need be only four inches, because the four-inch "veneer" is included in the eight-inch total requirement.

TOOLS

Many of the tools used in bricklaying are the same as those used for concrete block. They are listed here along with any differences. For further description of some of these tools see Chapter 10.

trowel—basically the same tool as for blockwork, but perhaps even more important here because it is used more intensively. Strength can be particularly vital, since experienced bricklayers will use the edge of the trowel to cut a brick in preference to the bolster or chisel peen of the hammer. The 10-inch size is usually recommended for brickwork, although the beginner may feel more comfortable with a smaller one. Usually, the longer the trowel, the easier it is to use, but a long trowel requires a strong wrist because the mortar is placed farther away from the handle.

bolster—the same blocking chisel used for blockwork, except that a smaller one is better for brick.

mason's level—same tool as for blockwork.

six-foot rule—same tool as for blockwork. You might use the spacing rule more often with brick.

line and pins—same tool as for blockwork.

story pole—used to insure that all courses are of equal height, particularly where two sides are built separately over an opening. The pole is usually a piece of 1 × 2 marked with course heights on three sides and with lintel, frame, and sill heights on the other. It is generally 10 feet (one "story") high.

gauge pole—a smaller story pole used to check uniform thickness of courses and mortar bed joints.

Table 15-3. Estimating Brick Requirements

Area	Walls (Single)	Paying Brick
1 sq. ft.	6.16	4.5
5 sq. ft.	30.80	22.5
10 sq. ft.	61.60	45.0
25 sq. ft.	154.00	112.5
50 sq. ft.	308.00	225.0
75 sq. ft.	462.00	337.5
100 sq. ft.	616.00	450.0
Add 5 to 10 percent for waste and breakage.		

ORDERING BRICK

Unlike concrete block, brick comes in only one size and shape per type. There are no half-bricks or special bricks for corners, headers, or anything else. This may make construction more difficult, but it makes ordering quite simple. A rather accurate guide to use is seven common bricks per square foot of wall layout, or 4 1/2 bricks per foot of horizontal installations like a patio. (See Table 15-3). Brick can be purchased at all masonry and most building supply dealers.

221

16

Horizontal Techniques

Horizontal bricklaying is not really bricklaying in the true sense of the word. It simply involves laying brick in a bed of sand or mortar. As such, it is an excellent way for the novice bricklayer to get accustomed to working with brick and, at the same time, to enhance the usefulness and beauty of his home.

As Figs. 16-1 and 16-2 show, a brick walkway or patio is a handsome addition to any home. This type of installation is simple to make and relatively inexpensive. Rustic-looking and trouble-free, it is highly recommended as a beginning brick project.

BRICK-IN-SAND

It is easier, if a little less permanent, to lay brick in a bed of sand than in a bed or mortar. This mortarless paving technique does not require a concrete slab base, as does the mortared method, and it can be executed much more quickly (Fig. 16-3). In some regions, the brick may be subject to heaving from frost, but any damage can be easily rectified by simply resetting the bricks that have been disturbed.

In the brick-in-sand method, sand is used both for a base and for filling between the cracks. The sand does a surprisingly effective job of holding the bricks in place if properly used. The job is even simpler if you have naturally

Fig. 16-1. Brick terraces have an attractive, rustic appearance.

sandy soil, which was the case in the Long Island patio shown in Fig. 16-3. Those who live in regions without sandy soil will have to import a truckload of sand to use this method, but sand is cheap and easy to find.

Edging can be made with more brick, set on end in the *sailor* position and staked every four feet, unobtrusively. (See brick position nomenclature in Fig. 19-2.) For an even more rustic look, use railroad ties for edging and stairs. Ties used to

Fig. 16-2. Brick walkways can also be very attractive and rustic-looking.

Fig. 16-3. A brick-in-sand patio is easy to build. Use the back of a rake to level the sand bed.

be cheap, but no so anymore. They are still reasonably priced, however, in areas where track is being torn up. Ties are extremely heavy and hard to move around, but one or two of them are usually all you will need for each edge, and ties can be cut into two or three lighter sections to serve as stairs. Make sure the ties are treated with preservative to prevent rotting and termite infestation.

GETTING THE MATERIALS

The easiest brick to use for patios and paths is *paving* brick, which is exactly twice as long as it is wide, as discussed in the previous chapter. If you use *common* brick in any patterns other than #1 and #3 in the top row of Fig. 16-4, you must leave spaces to compensate for its slightly irregular dimensions, and your layout can go awry quickly. Stay away from all other brick types except *paving* brick in order to keep the layout simple.

Used brick was quite inexpensive at one time, but now it is almost as costly as new brick, and can be hard to find. Furthermore, the old mortar can get in the way of pattern work, making it hard to manipulate for this type of brickwork. Figure on using five standard or common bricks per square foot, although your actual needs will be a little less than this

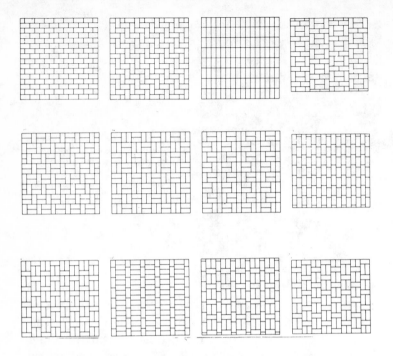

Fig. 16-4. Some of the types of bond used in brick paving. For common, non-modular brick without mortar, the first and third patterns in the top row are the easiest to construct.

figure. Five will allow for about 10 percent waste and cutting. If you are using paving brick instead of ties around the perimeter, figure three bricks per liner foot. Add a few extra if common brick is being used because it is 1/4 inch shorter in the *sailor* or on-edge position.

The best type of railroad ties are pressure-treated with creosote and will last almost forever. Actually, if the ties have really been used in railbeds they have taken severe punishment, so look them over on all sides to see if they are still in good shape. New railroad ties sold at lumberyards are usually 6 × 8s with a coat of creosote on the surface. They are not as durable as true used ties, but they are considerably lighter and easier to cut. If you are using real used ties, try to plan your patio in exact lengths of the ties (8 to 8 1/2 feet). The pressure-treated wood can be cut only with a chain saw, and the heavy creosote oil raises the very devil with the blade.

For ties, you should have some very long galvanized nails—the biggest you can find, at least 20 pennyweight. Use the same *mash* hammer you used with the bolster to drive these spikes. If you can't plan so that pressure-treated ties will be used full-length, and you must cut them, a chain-saw is essential. Chain-saw rentals cost about $15 a half day, so avoid that route if at all possible. Non-pressurized ties can be cut all around with a regular power saw, then finished off in the middle, if necessary, with a hand saw.

LAYOUT

One advantage of the tie-bordered walk or patio is that it can be located on sloping ground. Ties can be built up on the lower side, but don't use more than two or three in height. Otherwise, you'll be making a true retaining wall, which is a lot more complicated. Also, you'll need a railing to prevent accidents.

The size of your project is determined by your needs and available space, but also try to work your layout so that you do as little cutting of brick and ties as possible. Don't insist on an 18-foot patio if the ties are 8 1/2 feet long. Settle for a 17-footer. When you have all your dimensions worked out, lay out the periphery with lime. Set stakes to the same height at each corner, using a line level and cord. Although it helps to be as accurate as possible, slight gradations are not particularly objectionable with this type of rustic brickwork. If you are using mortar for construction, you should allow a slight grade of 1/4 inch to each foot for drainage. Drainage is no problem with the mortarless type of construction.

Setting the Ties

After you have determined the exact layout, dig down from the intended top surface to a depth of six inches all along the perimeter. Your trench should be eight inches wide, at least, but don't worry about going over on the inside. You will be digging that out next for the brick. Where the ground slopes, deposit the earth in a low spot. If necessary, build up low patio sides with more ties, toenailing (fastening with toed nails) every few feet with your long galvanized spikes. Toenail each

Fig. 16-5. Brick can usually be cut in half with a single blow with the bolster and hammer. Wear goggles for eye protection.

tie together at the ends, too, to keep them from moving (although movement is not too likely in any case, as you can tell when you start lugging them around).

Once the ties are in place, level them by filling in low spots with earth or sand. No finish is necessary except where there might be a big gouge in a new tie. In that case give the bare spot a shot of creosote or similar preservative.

Setting the Field Brick

When you are finished laying the railroad ties, you are ready for the brick. Where the soil is already sandy, dig down to the thickness of the brick plus a little bit extra for leveling. For common brick, 2 1/2 inches is about right. Where sand must be imported, dig down about 4 1/4 inches, leaving room for a two-inch sand bed. (Brick laid on a concrete slab will be discussed in a subsequent section).

Lay your sand bed, if necessary, and rake the surface level. No matter how hard you try, you'll have to individually level each brick, but it is easier if you get the subgrade as level as possible. For a perfectly level surface, use a *screed board* as described for concrete.

Start your pattern in one corner. Lay down a whole module at one time, leveling the individual bricks by pushing sand underneath or scraping it off where bricks are too high. You can use a carpenter's or mason's level for this, but you will probably do just as well eyeballing it and checking every so often with the level to make sure you aren't too far wrong.

If you're using running bond, which is the easiest pattern in the long run, you'll have to cut half bricks for every other course. To do this, place the brick in the sand, hold the bolster in the middle, and give it a solid, quick rap with your hammer (Fig. 16-5). The brick will break pretty much in two, although your line probably won't be nice and neat. If you are fussy, score the brick all around before whacking it, but little imperfections, again, are tolerated—even welcomed—in this type of brickwork.

For a large patio, run crossing string lines between the corners to make sure you don't get too far off as you approach the middle. Don't be too concerned if there are slight rises or hollows. Again, it's a rustic type of thing you're doing here.

LOCKING

All your bricks should have been laid as closely together as possible. But each brick is not a perfect rectangle and you aren't the perfect bricklayer. There will be gaps between bricks, some fairly large, most quite small. To make sure that all the bricks are locked tightly together, spread sand over the whole surface and work the grains down between the bricks with a garage broom (Fig. 16-6). Keep working the broom over

Fig. 16-6. To lock bricks in place and fill in gaps beneath them, spread sand over the surface and brush it down between the cracks.

and over the cracks, and eventually the sand will all disappear. You may wonder where it all goes; a lot of sand will filter under as well as between the bricks.

To complete the *locking* process, water the bricks gently but thoroughly, and watch the sand disappear even further. Spread some more sand around, work it down in again, and rewater. You will probably find that this step is necessary three or more times in all.

STEPS

There are lots of ways to build steps, and the method chosen depends on the topography as well as your taste. Brick-and-tie walkways, though, go particularly well with brick-and-tie patios, as shown in Fig. 16-7.

For a main entrance to the house, you may want to use full ties, but halves and even thirds are wide enough for most walks. Don't be concerned if some sections are longer than others: this will occur because the ties are not always the same length. You will use them all someplace.

Used ties can be put right to work, but the cut edges of new ties will need some creosote, so apply this and let it dry before laying new ties. You will need two sections of ties for each step because one tie is not quite wide enough to get a good foothold.

Fig. 16-7. Brick-and-tie stairs match the patio above them.

In work like this, the conventional wisdom dictates to plan ahead of time, but it is easier to play it by ear as you go along when dealing with tricky topography. Figure out where you want to start and where you want to end; the remainder of the work is easier done than planned. Keep laying steps until you reach a point where it is easier to lay sloping or level walk, then continue until the grade gets too steep for walking. Lay a few more steps, then a walkway, etc. It is best to put in the steps first and let the walks go until later.

The two-tie sections for each step should be toenailed together and, if possible, the top of each step should be nailed to the bottom of the one above. In addition to the spikes, though, earth or sand should be packed in against the edges, so that the steps won't be going anywhere even if you can't manage to nail them all together. In problem areas, set stakes and nail them to the ties.

WALKWAYS

You can use ties to edge your walkways if the walk is fairly wide, but the eight-inch-wide ties look out of scale with narrower walkways. In most cases, creosoted 2 × 4s or bricks set on end look better. Match the edging here, if possible, with whatever you use for a patio.

You will probably have to creosote the 2 × 4s yourself, so buy yourself a large can of preservative and a cheap brush and go to work. Pure creosote, if you can find it, should be used out-of-doors; be careful not to get any in your eyes. Creosote may even bother some people's skin, so wash any spills off as soon as possible. Be sure to get the ends of the 2 × 4s after they are cut, too. Nail the 2 × 4s between each step at the ends of the walks.

After the 2 × 4s are set in place, lay your brick in the same pattern as the patio. You will find it a different process to achieve the same modular patterns in some walks, however, so it may be necessary to change to a running bond for the walkways. No matter what the pattern, you may wind up cutting brick lengthwise as well as crosswise. If you do, it is wisest to first score the brick all around, because the bolster will not reach all the way across the mark in one blow.

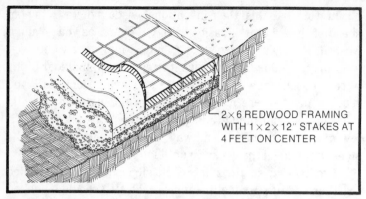

Fig. 16-8. For best results using mortared paving, lay a four-inch concrete slab, and top it with a 1/2-inch bed of mortar. Note the use here of 2×6 redwood for edging (Courtesy BIA).

When all the bricks are set, lock them in position with sand and water as above.

MORTARED BRICK

For those who prefer neat and uniform surfaces, frost problems can be eliminated by using mortared brick on a concrete slab (Fig. 16-8). Excavate the patio or walk area to a depth of at least six inches below the frost line, then fill with gravel to a height approximately 6 3/4 inches below the finished patio surface. Lay a concrete slab (as described in Chapters 1—7) approximately four inches thick. When the concrete has cured, lay a bed of mortar 1/2 inch thick over the area to be worked. Using common brick, butter all sides of each brick with mortar and place into the mortar bed with 1/2 inch joints in between. Make a dry run first, however, in a small section, to make sure that all mortar joints will be the same. If not, cheat a little to create the proper pattern. (Exact mortar joints are unimportant in this type of construction.)

This method is much more difficult and time-consuming than brick-in-sand work, but the owner has the pleasure of knowing that his patio will be neat and level, and will stay that way for a long time. Also, the attractive herringbone, bas-ketweave, and other modular patterns can be accomplished without great difficulty, using uniform or slightly variable mortar joints.

17

Building Walls

When you build a brick wall or other three-dimensional structure, you have to know what you are doing. If you do a poor job of laying a brick patio, the worst that can happen is that you have a somewhat unsightly patio. If you do a poor job of building a wall, it can come crashing down on your head—or someone else's.

But this is not meant to frighten you. No one is suggesting that you build a cathedral the first time around. Ideally, you should, in fact, begin with a small, single-wythe (tier) wall in order to learn the proper technique. As you advance, you may tackle the harder projects, but first things first. Even if you get fairly proficient, you may never make it as a professional bricklayer. To do this type of work professionally demands lots of hard practice, and—more important to any employer—the ability to put up a lot of brick quickly. Basic, elementary bricklaying is emphasized here. Tricks to speed up the job and achieve proficiency are described later.

MIXING THE MORTAR

Begin by thoroughly blending the cement, lime, and sand (as described in Chapter 15) with your shovel. Then scoop out a hollow in the middle of the mixture and slowly add water until it is the consistency of soft mud. Unlike concrete and

Fig. 17-1. Wet brick about an hour before laying it, so that the surface is dry, but the interior stays damp (Courtesy Louisville Cement Co.).

some other mortar mixes, it is better in this case to err on the side of too much water than too little, since the brick will absorb much of the moisture anyway.

Make sure that there are no dry spots in the batch, and mix only as much as you can use in 30 minutes to an hour. To prevent too-rapid drying out of the mortar, wet the brick with the garden hose about an hour before using (Fig. 17-1). The surface of the brick should be dry when laying up, so that the mortar does not slip off, but the interior will still be wet and will help avoid premature mortar stiffening. It also helps to dampen the foundation before laying the first course.

BUILDING A SIMPLE ONE-TIER WALL

Brick is heavy material, and when combined with mortar it weighs about 120 pounds per cubic foot. Obviously, a brick structure cannot stand upon a weak foundation. A brick barbecue, for example, should rest on reinforced concrete. Any brick wall should bear on concrete footings at least twice as wide as the brick itself (see Chapter 8). Smaller, less massive structures like garden edging can be set in mortar on top of a sand bed, but most brickwork demands at least a four-inch slab underneath.

Brick patterns are called *bond*, and there are many choices, as illustrated in Fig. 17-2. The most common and easiest to lay is *running bond*, in which each brick is placed exactly halfway over the tops of the two bricks below it, half on each side of the joint. The other bonds are used mostly in double-wall or wythe construction. No matter which bond you choose, lay out the first course in a dry run so that you wind up with a whole or half brick at each end. You can usually avoid cutting brick by adjusting the length of the row a little or, if necessary, using tighter mortar joints (as little as 1/4 inch is acceptable).

Whether the structure is one or two bricks wide depends upon the particular project. A high wall should be two tiers wide for extra strength, but there is no need to use more than a single wythe for a small project like a planter.

Begin laying brick at a corner. Build it up three or four rows high, dovetailed with the adjacent wall (if there is one) at right angles. Make frequent use of the level and square to keep things in true. When you have built up one corner, go on to the next, lining it up with the first by means of a straightedge or line level.

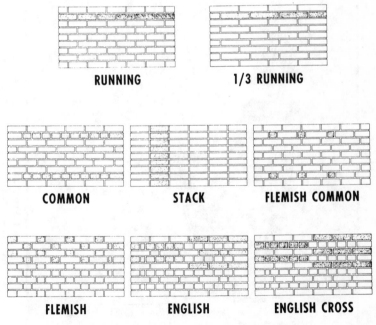

RUNNING

1/3 RUNNING

COMMON

STACK

FLEMISH COMMON

FLEMISH

ENGLISH

ENGLISH CROSS

Fig. 17-2. Some of the more typical types of brick bond.

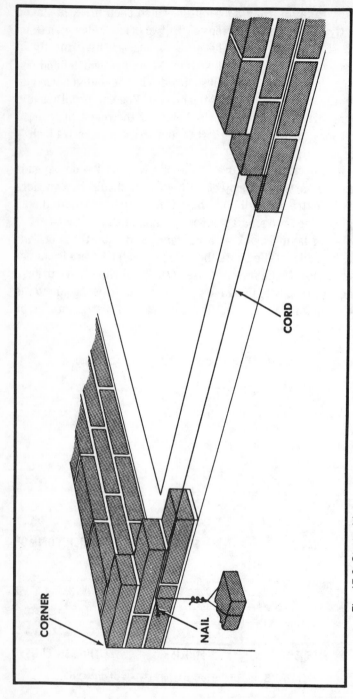

Fig. 17-3. Start with two corners, after making a dry run, then drive nails or mason's pins into mortar and run a line between each end of the first row.

CORD

CORNER

NAIL

With at least two corners in place, push nails or mason's pins into the soft mortar between the first two courses or rows and run a cord between them (Fig. 17-3). The line or cord should be almost, but not quite, touching the front and top of the first course. Use this string as a guide for placement of the rest of the bricks in the first row. Each brick should be lined up the same distance from the line. You should be able to do this by eyeballing, but it may help if you measure the first few bricks. After the first course is laid, move the string up to the row above and repeat for each course.

The technique of bricklaying is simple, as long as you take your time in the beginning. Take a slice of mortar on your trowel and lay it along the foundation about a half-inch thick and long enough for two or three bricks—later you can stretch to four or five (Fig. 17-4). Furrow the mortar a little bit—about 1/4 inch—with the point of the trowel, making sure the mortar extends the width of a brick (Fig. 17-5). Then *butter* the end of a brick and set it firmly into the mortar bed. Some mortar should be squeezed out of the joints. Trim off the excess mortar with the side of the trowel and return it to the mortar board. Butter the next brick, set it against the previous one, and continue. With the first few joints, you may want to measure the mortar (half-inch thick is best), but you will soon learn to eyeball this, too.

If you've made your dry run correctly, the last closure brick in the center should fit perfectly. When laying this brick,

Fig. 17-4. Lay down a 1/2-inch bed of mortar on the foundation or on the brick below.

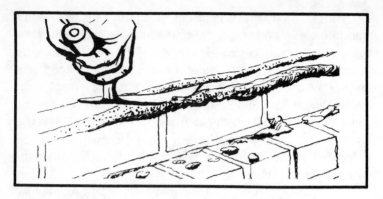

Fig. 17-5. The mortar is furrowed down the center, about 1/4 inch deep (Courtesy BIA).

butter *both* ends of the brick as well as the flanking bricks on either side. Lower this brick carefully into the gap, taking care not to knock off the mortar, and tap it into position. If the brick is too big, cut it as described in the previous chapter. If it is too short, you can cheat a quarter inch with extra mortar. To fill any more than that, fill in with a piece of brick—or knock it down and start over.

Any brick that is set too high can be tapped down into position, but you can't tap a brick that has been set too low. Pull up the brick, scrape off the mortar, lay down more mortar, and try again.

Keep working on the intermediate bricks between corners until you reach the highest course, then switch to laying corners for three or four more courses. Go back to the in-betweens, then the corners, and so on, until you reach the top.

Once You're Underway

As you move along, don't forget to keep checking for plumb, level, and alignment. If you find your wall is bulging, dipping, or otherwise running out of true, don't attempt to tap bricks into place once the mortar sets up. Take up the brick (and those on top, if necessary), scrape off the old mortar, add new mortar, then reset it. Any tampering with the mortar once it starts to harden is sure to cause cracks or hollows in the joint, weakening the entire structure and inviting leakage.

As each course is completed, and before the mortar has hardened, run your 5/8-inch pipe or jointing tool across the joints to produce a smooth and uniform appearance. Other types of tooling are often used, but the concave joint produced by the pipe is the best for producing a water-tight seal. Other joints are illustrated in Fig. 17-6. While tooling (or *pointing*), also fill in the holes left by your line nails, and any other imperfections, with dabs of mortar.

Finishing Touches

Like all cement compounds, mortar should be allowed to mature for a couple of days at least. Hot, dry weather will accelerate the drying process, and make it proceed too quickly. During such days, give the wall an occassional

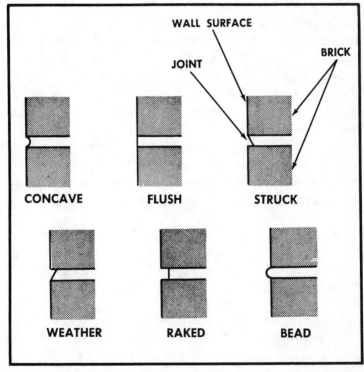

Fig. 17-6. The most commonly used joints in brickwork. The "weather" joint is most weather-resistant, but difficult to form because the trowel must be used from underneath. The concave joint is almost as weather-tight and much easier to tool. Other joints are not as water-resistant.

watering. And be gentle to your new masterpiece for a week or two, until the mortar is good and strong.

Meanwhile, clean up any mortar stains with damp burlap. If this won't do the trick because the stains have set, remove them with a solution of muriatic acid diluted with 10 parts of water. Wet down the brick and scrub on the liquid with a stiff brush. Wear gloves and goggles when using this solution, and splash off any acid immediately and thoroughly with cold water.

After the mortar has completely cured (at least a week), you may want to coat the brickwork with a colorless masonry sealer. This provides a transparent, water-repellent surface that protects against crumbling mortar for many years.

LARGER AND STRONGER WALLS

While the basic bricklaying techniques for small and large walls are the same, the one-wythe wall described previously is primarily decorative, and should not be constructed as a load-bearing wall or as one that will hold back great forces, as in a retaining wall. Nor should it be built too high. Four feet is maximum for low-wind areas and as little as two feet should be considered the limit where there are high winds. Larger walls can be blown over or toppled by a good hard shove. If you build a single-wythe garden wall higher than four feet, it should

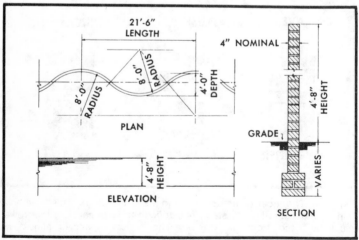

Fig. 17-7. The plan for a serpentine wall—rugged and good-looking.

follow a sepentine pattern as shown in Fig. 17-7. The curved lines offer more resistance than would a uniform flat surface. Straight walls higher than eight feet should be supported by pilasters on a firm foundation.

There are many types of larger walls; the common property of all of them is greater strength and more durability than a single-wythe wall. Sometimes a double brick wall is built, a single brick wall is provided with other backup materials such as concrete block, wood, or structural clay tile. Building codes should be meticulously followed when building a wall of this type.

Types of Load-Bearing Walls

There are basically four types of walls which have the strength and durability to be used as load-bearing walls. These are:

solid masonry: A solid brick masonry wall consists of two walls bound closely together. The two wythes are built one next to the other and locked together either with metal ties or by structural bonding. Structural bonding is created by using brick patterns in which the brick itself connects the two walls. Examples are common, English, and Flemish bonds. *Common Bond* is a pattern formed when every fifth to seventh course (depending on strength requirements) consists entirely of headers. In *English Bond*, the courses consist of alternate stretchers and headers. *Flemish Bond* alternates headers and stretchers in every course, arranged in a way that the headers and stretchers in every other course appear in vertical lines (Fig. 17-8). These overlapping bonds develop great structural strength because the stretchers have longitudinal strength and the headers bind the wall transversely. Most building codes require that structurally bonded brick walls be built so that no less than four percent of the wall surface is composed of full headers, with the distance between adjacent headers no more than 24 inches either vertically or horizontally.

Structural bonding can be created with brick plus other types of masonry such as block or clay tile, but the patterns here are intricate and difficult for the handyman. The English or Flemish bonds described are easier to build and are

241

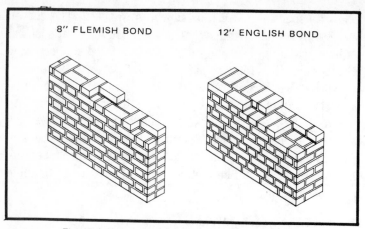

8" FLEMISH BOND   12" ENGLISH BOND

Fig. 17-8. Examples of solid masonry construction.

handsome to look at, but neither of them is recommended as a beginning project. Common bond is the easiest of all, but still requires exacting workmanship. It is hard enough for the novice to get one wall perfect, much less two at once; but once you have mastered the basics, these walls are an interesting next step up the ladder.

If you want to build a solid masonry wall, one relatively simple method for the beginner is to use metal ties with concrete backerblock. Use rigid steel ties at least 3/16 inch in diameter, spaced so that at least one tie occurs for every 4 1/2 square feet of wall area. Maximum vertical distance between ties is 18 inches, and horizontal maximums are 36 inches. The ties in alternate courses must be staggered.

cavity wall: A cavity wall is made up of two wythes of brick or one of brick and the other of block or clay tile. The wythes are connected with metal ties, but with a space in between (Fig. 17-9). The space is usually filled with poured insulation. A cavity wall is built much like the solid masonry wall described above. Here, however, the walls are built separately with about four inches between them. As the walls are built up, water-repellent vermiculite or perlite is poured into the cavity.

A variation of the cavity wall is the masonry-bonded hollow wall (Fig. 17-10). This works on the same principle, except that 4 × 4 × 12-inch bricks are laid in a locking

242

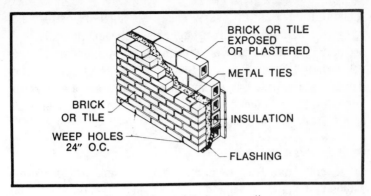

Fig. 17-9. A metal-tied cavity wall.

pattern, with headers bridging the gap. These are placed two feet on center horizontally and from 20 to 24 inches vertically.

reinforced masonry: This type of wall uses steel reinforcing bars set into the footings and sometimes horizontally (Fig. 17-11). The wythes are built on either side of the rebars, and the cavity is filled with grout. This is an extremely strong wall and is often used for retaining walls or other structures where great resistance is required. Care must be taken not to allow too much mortar to drop into the cavity and interfere with grouting. For this reason, bed joints must be carefully laid and leveled away from the cavity.

brick veneer: In this instance, the brick masonry itself plays no part in the bearing of the wall. The wood or block wall is the only part that counts structurally, and it must be built to hold up by itself without the brick veneer. Codes should be

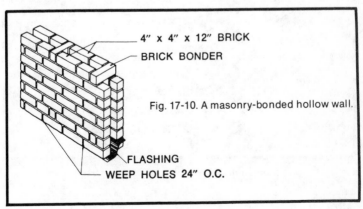

Fig. 17-10. A masonry-bonded hollow wall.

consulted on this in every case. Corrugated metal ties, 22-gauge thick, are usually used for ties between the walls, and are nailed to the studs or block with 8 pennyweight nails. Weep holes should be provided in the brick course just above the flashing at the base by leaving out head joints about every two feet (Fig. 17-12). Most codes limit the height of the veneer to 20 or 30 feet above the foundation.

Some Professional Tips

If you have any thoughts about turning pro or doing extensive brickwork around the home, here are some tricks of the trade that will make your work faster and easier:

A common waste of time in laying brick is the failure to use both hands as much as possible. Right-handed bricklayers have a tendency, unless they have trained themselves otherwise, to use their left hand only intermittently. The opposite is true of lefties. Yet it is not at all difficult to train yourself to use both hands nearly all the time. When you do this, the result is a faster rate of bricklaying, less physical strain, and no decrease in the quality of workmanship.

Learn, for example, to pick up a brick from the pile and a trowelful of mortar at the same time.

When working on a double-wythe wall, it also saves time to *wall* the brick—that is, to place it on the built-up section within

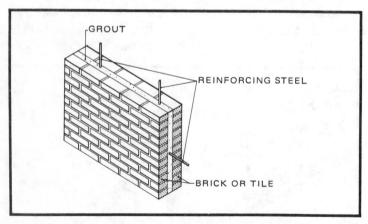

Fig. 17-11. A reinforced masonry wall, with steel re-bars set into footings. Horizontal reinforcement may or may not be required, depending on codes and local conditions.

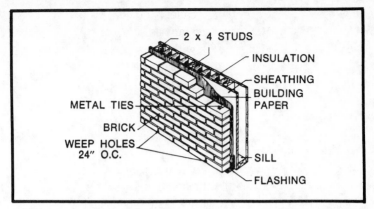

Fig. 17-12. In a brick veneer wall, although it is very common and attractive, the brick does not ordinarily count as part of the structure. Code regulations are important here.

easy reach. Obvious though it might sound, remember to use both hands to pick bricks from the stock pile and place them on the wall.

After the bricks are *walled*, the handyman (and some pros) will usually pick up one brick and lay it in the mortar, then cut off the excess mortar and place his head joint. After this, he will pick up another brick and lay it, placing a full head joint. By doing this, however, one hand is idle while the otherhand is cutting off mortar and placing a head joint. You can easily train yourself to put your idle hand to work by picking up another brick and having it ready to lay while the "busy" hand completes the placing of the head joint of the previously laid brick.

18

Brick Projects

The projects in this chapter are relatively easy, and can be built once a basic knowledge of bricklaying is learned.

OUTDOOR BARBECUE

Sand-finish brick is recommended for outdoor barbecues. It reduces cleanup work after the job because mortar won't stick or smear, although crumbs will still have to be brushed away.

Excavate and pour the concrete foundation to the dimensions indicated, placing reinforcing bars as shown in Fig. 18-1. Criss-cross the bars in a grid pattern and prop them up with brick units so that they lie approximately in the center. If you prefer, they can be wired together and handled as a unit.

Draw the outline of the barbecue on a foundation slab, leaving at least two inches all around. Lay out the first two courses of brick to see if the pattern works, allowing 1/2 inch where the mortar joints will be.

After a dry run, build the corners first, going three or four courses high, then filling in the wall from corner to corner. The bottom course should be bonded to the slab with mortar. Use a hand level frequently to keep the wall plumb and the rows of brick level.

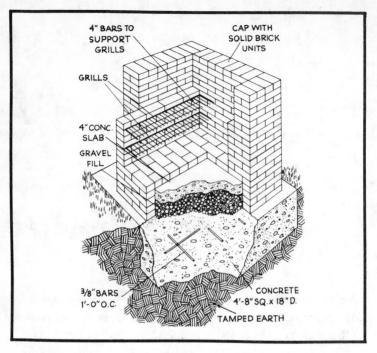

Within the figure:
4" BARS TO SUPPORT GRILLS
CAP WITH SOLID BRICK UNITS
GRILLS
4" CONC. SLAB
GRAVEL FILL
3/8" BARS 1'-0" O.C.
CONCRETE 4'-8" SQ. x 18" D.
TAMPED EARTH

Fig. 18-1. Construction details for an outdoor barbecue (Courtesy BIA).

Excess mortar may be clipped off with the trowel two or three rows at a time. As soon as the mortar is fairly hard, use a mason's pointing tool to shape and compress the mortar between brick.

As construction proceeds, insert four-inch lengths of reinforcing bars in the mortar joints to support the grills as shown. Use solid brick units for the top of the barbecue walls. Excess crumbs of mortar which remain when the wall is finished may be brushed away with a soft fiber hand brush.

MORTARLESS BARBECUE

You should be able to build a mortarless barbecue in a couple of hours. Bear in mind one important feature of mortarless brick work: the site you select must be absolutely level and stable. A concrete slab is best.

Buy your grill racks first; sometimes the availability of certain sizes is limited. You can adjust the construction of the barbecue to fit the grill racks you buy.

As you follow the pattern shown in Fig. 18-2, be careful to tap in the brick units that support the grill racks before proceeding to the adjacent work surface. The brick units supporting the grill racks should be tapped in with a 2 × 4 and a hammer. Lay the 2 × 4 edgewise to the row as you tap the bricks.

SCREEN WALL

An attractive brick screen will dress up your yard and keep trash cans, and other potential eyesores out of sight (Fig. 18-3). It also lets breezes through.

Some local building regulations do not allow this kind of brick screen; check your municipality's ordinances. If such screens are allowed, ask how deep the concrete footing must be according to the local codes.

Study Fig. 18-4 carefully; this is not a project to rush into. Have some experience behind you first on a less complicated project.

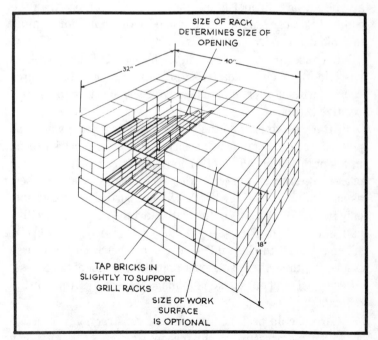

Fig. 18-2. Construction details for outdoor barbecue without mortar (Courtesy BIA).

Fig. 18-3. Screen built from brick. (Courtesy BIA).

Excavate the footing, making sure it goes below the frost line and that the bottom of the trench is as level as possible. Check it with a straight 2 × 4, 8 feet long, and check also with a two-foot or four-foot hand level. Dig out no more than the width of footing wanted. Use the sides of the trench for forming to avoid using extra materials.

Because the screen's corners must be built as plumb as possible, it's a good idea to begin by laying out the first three courses without mortar. This will give you a chance to insure a positive layout.

After establishing the solid wall portion, start the pierced-pattern portion at the corner by cutting a brick to a six inch length (A) in Fig. 18-4B. Then cut a brick in half and lay in place, centered on the joint where two bricks come together, as shown in (B) of Fig. 18-4B. The cut portion will be visible only from the back side of the wall. Start the next course with a half unit, (C) in Fig. 18-4B, followed by a whole brick (D) spanning over to the middle of the cut brick on the course below. Continue these stretcher units for the rest of the course. After the stretcher course is in place, repeat the first pattern described.

Care should be taken during the construction process to avoid putting pressure on the screen. Remember that it's only a little over half as heavy as a solid wall.

BRICK SANDBOX

A brick sandbox in your back yard will give your children a clean place to have fun. It is not particularly difficult to construct, either.

Dig an excavation for the sandbox as indicated in Fig. 18-5B and install a wooden frame. Fill the sandbox with sand and surround with a 1 × 4 edging as indicated. Spread a dry mixture of three parts sand to one part cement over the ground area around the sandbox, and tamp and grade as necessary to achieve a smooth, even surface. Place solid brick units directly upon this cushion base.

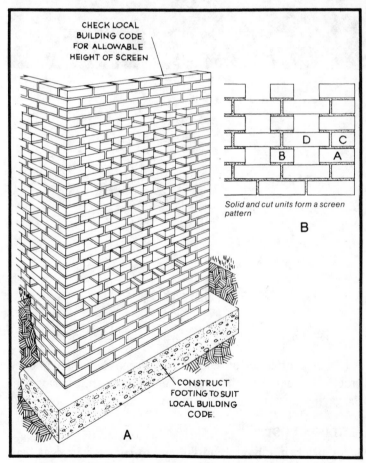

CHECK LOCAL BUILDING CODE FOR ALLOWABLE HEIGHT OF SCREEN

Solid and cut units form a screen pattern

B

CONSTRUCT FOOTING TO SUIT LOCAL BUILDING CODE.

A

Fig. 18-4. Construction details for brick screen (Courtesy BIA).

A

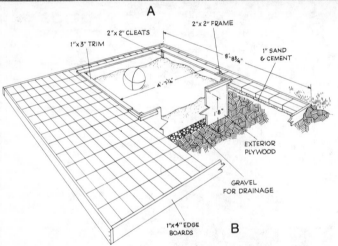

1"x3" TRIM
2"x2" CLEATS
2"x2" FRAME
1" SAND & CEMENT
8'-8¾"
4'-7¼"
8"
EXTERIOR PLYWOOD
GRAVEL FOR DRAINAGE
1"x4" EDGE BOARDS
B

Fig. 18-5. Photo and construction details for sandbox (Courtesy BIA).

Upon completion of the sandbox, sweep clean sand into the cracks between the brick units. Clean with a gentle spray from the garden hose.

MAILBOX POST

A brick mailbox stand is a more complex project than those described so far, but it offers the true handyman a

252

chance to prove his skill. The finished product will add a unique touch to suburban or rural homes.

An excavation 16″ × 24″ will be needed (Fig. 18-6). Choose a site where the earth has not been disturbed; if the mailbox is sited in fill material, the masonry will eventually settle out of plumb. Check local building regulations to make sure that the bottom of your concrete footing is below the frost line.

The mailbox will be built in an inverted position—that is, from the top down—so you will need to choose a level, convenient work area out of the family's way. This work area should be as close as possible to the site selected for the mailbox because the brick posts are going to weigh about 130 pounds each! It will be easier to work on an existing concrete sidewalk or driveway, or on a heavy piece of plywood leveled and stabilized.

Place the first three brick units as indicated on the top of Fig. 18-6B. In the beginning, because only the mortar holds these units together, you may want to make a small outline with thin wood strips to hold the bricks in place. Next, fill a core with mortar and insert one of the four inch lengths of 1/4-inch bar shown in Fig. 18-6. Then cut a brick in half and lay

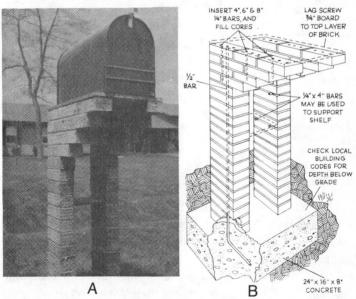

Fig. 18-6. Photo and construction details for mailbox post (Courtesy BIA).

Table 18-1. Materials Lists

OUTDOOR BARBECUE

450 cored brick units 3 3/4″ by 2 1/4″ by 8″
75 solid brick units, same dimensions
6 cubic feet of mortar
27 cubic feet of concrete for foundation
1 2/3 cubic feet of concrete for hearthslab
20 reinforcing bars 3/8 inch in diameter:

    5 bars 18 inches long
    3 bars 32 inches long
    12 bars 4 inches long

MORTARLESS BARBECUE

236 solid brick units 3 3/4″ by 2 1/4″ by 8″
2 grill racks

SCREEN WALL

Solid brick units 3 3/4″ by 2 1/4″ by 8″, 4.4 units per square foot of screen; 6.55 units per square foot of solid wall (allow 5% to 25% for waste)

    .90 cubic feet of concrete per lineal foot for 8″ × 16″ footing
    .66 cubic feet of concrete per lineal foot for 8″ × 12″ footing
    mortar as needed

SANDBOX

240 solid brick units 3 3/4″ by 2 1/4″ by 8″ or 4″ by 8″ by 1 5/8″ and about 5 cubic feet of damp, loose sand (1/4) ton) (does not include sand for sandbox)

1 1/4 cubic feet of cement or 1 1/4 bags of portland cement galvanized nails

5/8 inch, exterior grade, rough-sawn plywood siding, coated with preservative, cut into:

    4 lengths of 20″ × 55 3/8″
    4 lengths of 2″ × 2″ × 7′
    4 lengths of 1″ × 4″ × 9′
    2 lengths of 1″ × 3″ × 10′

MAILBOX POST

60 cored brick units 3 3/4″ by 2 1/4″ by 8″ (includes allowance for waste)

1/4 inch pencil rod reinforcing cut as follows:

    2 lengths of 4″
    2 lengths of 6″
    4 lengths of 8″

2 lengths of 1/4 inch reinforcing bar 53 inches long
5.3 cubic feet of concrete
mortar as needed

it in place along with two full-length units. Fill only those cores which will receive six-inch and eight-inch lengths, as shown. Lay two full units, fill the cores, and insert two eight-inch lengths.

Next, cut a brick in half and lay a whole unit and a half unit together. Fill the cores and insert the 1/2-inch rod. It's a good idea to let this set before going further; if the rod moves too much, the mortar won't bond to it. Let it set for at least 2 1/2 hours.

Spread mortar and carefully thread the brick units over the 1/2-inch bar, until the indicated number of courses has been placed. Be careful to fill the cores completely each time a unit is placed. Drop each post on all four sides and allow it to set for about a week.

You are now ready to install the posts in a concrete base. Use brick units left over to support the post temporarily while the concrete is being poured. The bar sticking out of the end of each post should be bent at a 90° angle so that it clears the bottom of the excavation. Use a short piece of pipe or a pipe wrench to bend it. Set the posts in place and prop them in all four directions with 2 × 4 props. Also, prop the posts apart from each other with a 2 × 4 wedge. Use the hand level to make sure they are plumb. Pour the concrete and allow it to set for two weeks; cover the top of the posts during this period to keep water from getting into the masonry.

To install the mailbox, drive wooden plugs in the unfilled cores of the brick and leg screw a 3/4 inch plank to the plugs. This plank should be the size of the mailbox selected. Redwood is recommended for best weather resistance. Attach 1 × 2 redwood furring strips around the edge of the plank. Fit the mailbox over the furring strips so that they don't show. Nail the mailbox flange (this part projects below the bottom of most mailboxes) to the furring strips.

19

Specialized Brickwork Techniques

The work in this chapter is intended for the person who has already gotten his feet wet doing simpler brick projects. As a matter of fact, his feet should be *soaked* before he tries building arches, chimneys, and fireplaces. (Outdoor fireplaces or barbecues are an exception, and can be built successfully by a novice.)

Most of the projects described so far have been basic enough that the reasonably skilled handyman can build them with safety and assurance. Arches, however, are neither simple nor harmless. They demand exacting workmanship and are extremely dangerous if improperly built. No one wants an arch falling on his head. And, even if the arch misses the skull, its fall may cause the wall above to cave in. At best, severe structural damage will result.

An improperly designed or built fireplace can cause fires, smoke damage, and even death by asphyxiation. So approach this chapter with caution and a good track record of previous brick construction.

Now that you're properly frightened, it should be said that someone who's mastered the basics of brickwork can also master the techniques in this chapter with care and study. But do make sure you're doing it right.

Fig. 19-1. Using a "center" to build an arch.

MAJOR ARCHES

Arches are built with the aid of temporary shoring or *centering* (Fig. 19-1). Temporary centers should be adjusted with wood wedges. When the arch has set, the wedges are withdrawn and the center is released, insuring the safety of the masonry work.

The purpose of the centering is to carry the dead load of the arch and other loads until the arch itself has gained sufficient strength. In no case should centering be removed until it is certain that the masonry is capable of carrying all imposed loads.

For plain brick arches, it is recommended that centering remain in place at least ten days after the completion of the arch. Where loads are relatively light, or where the majority of the load will not be applied until some later date, it may be possible to remove the centering before the ten-day period has elapsed.

Before laying brick for an arch, the pattern should be carefully laid out so that you will have the proper number of courses, and uniform joints bonded on the face and on the *soffit* (the underside).

There are two general methods used in the construction of brick arches. One uses special brick shapes with mortar joints of constant thickness. The other uses units of uniform shape, and mortar thickness is varied to obtain the desired curvature. The method used is determined by the arch dimensions and by the appearance desired. In many instances, special shapes may be made by cutting standard rectangular brick units at the job site. Special manufactured shapes for brick arches may also be available. Check local sources.

The structural strength of an arch depends upon the compressive and bond strengths of the mortar used. Type M or type S mortar is recommended.

It is important that all mortar joints be completely filled in archwork. Brick arches are usually constructed so that, at the crown, the units will be laid in *soldier* bond or *rowlock header* bond (see Fig. 19-2). Under many circumstances, it is difficult to lay units in soldier bond and yet obtain completely filled mortar joints. This is especially true wherever the curvature of the arch is a short radius, and tapered or V-shaped mortar joints are used. In such cases the use of two or more rings of rowlock headers is recommended. Besides helping to achieve full mortar joints, rowlock headers provide a bond through the wall, thereby strengthening the arch.

When building a Roman arch, the *springing points* of the spring line (see Fig. 19-3) should be located about one or two

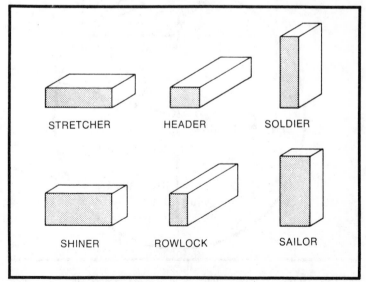

Fig. 19-2. Terminology for the various brick positions.

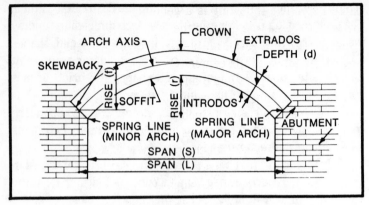

Fig. 19-3. Terminology for major arch construction.

inches below the radius point to correct the optical illusion of flatness, which a true semicircular arch will give. There are various methods of finding the *radius point*. Locating this point is necessary to properly lay out the curve of an arch, when the *span* and *rise* are known. (Arch terminology is explained in

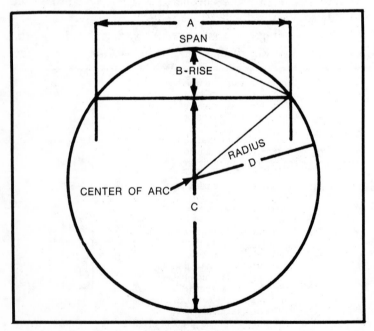

Fig. 19-4. To build a proper major arch, it is necessary to find the radius. Text explains the geometric method (Courtesy BIA).

Table 19-1). The following is a geometric method for finding the radius. It may be used on the job: Multiply one half of the span (A) by itself, (see Fig. 19-4), and divide by the rise (B) to find the distance to the center (C). To obtain the radius, (D), which is the objective, add B and C and divide by 2.

Example: To find the radius of an arch whose span is 48 inches and whose rise is 12 inches, multiply on half of the span (A) by itself (24 inches times 24 inches) to get 576 square inches. Now divide by (B), the rise, (12 inches). This yields 48 inches (C). This, (48) plus the rise (12) equals 60 inches; 60 divided by 2 gives 30 inches, which is the radius.

Table 19-1. Arch Terminology

> **abutment:** A **skewback** (see below) and the masonry which supports it.
>
> **arch:** A form of construction in which a number of units span an opening by transferring vertical loads laterally to adjacent units and thus to the supports.
>
> **arch axis:** The median line of the **arch ring**, halfway between the **extrados** and **intrados**.
>
> **chamber:** The relatively small rise of a **jack arch**.
>
> **crown:** The apex or top point of the arch ring. In symmetrical arches the crown is at mid-span.
>
> **depth:** The depth (D) of any arch is the dimension which is perpendicular to the tangent of the axis. The depth of a jack arch is taken to be its greatest vertical dimension.
>
> **extrados:** The convex curve which bounds the upper extremities of the arch.
>
> **intrados:** The concave curve which bounds the lower extremities of the arch (see **soffit**). The distinction between **soffit** and **intrados** is that the intrados is a linear distance while the soffit is the measure of a surface area.
>
> **jack arch:** A flat or nearly flat arch.
>
> **keystone:** The center unit that wedges an arch in place.
>
> **multicentered arch:** An arch whose curve consists of several arcs of circles which are normally tangent at their intersections.
>
> **rise:** The rise (R) of a minor arch is the maximum height of arch **soffit** above the level of its **spring line**. The rise (F) of a major arch is the maximum height of arch axis above its spring line.
>
> **segmental arch:** An arch whose curve is circular but less than a semicircle.
>
> **skewback:** The inclined surface on which the arch joins the supporting wall. For **jack arches** the skewback is indicated by a horizontal dimension (K).
>
> **soffit:** The undersurface of an arch.
>
> **span:** The horizontal dimension between abutments. For minor-arch calculations the clear span (S) of the opening is used. For a major arch, the span (L) is the distance between the ends of the arch axis at the skewback.
>
> **spring line:** For minor arches, the line where the **skewback** cuts the **soffit**. For major arches, the term commonly refers to the intersection of the **arch axis** with the **skewback**.

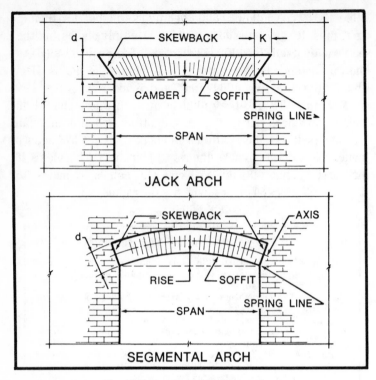

Fig. 19-5. Two types of minor arches.

MINOR ARCHES

Figure 19-5 shows a *segmental arch* and a *jack arch*, which are often used to span openings not exceeding six feet in width. The *rise* of the segmental arch should be not less than one inch per foot of the *span*, and the *skewback* should be at right angles to the arch axis. In the jack arch, the *camber* should be 1/8 inch per foot of span and the *inclination* of the skewback, K, should be 1/2 inch per foot of span for each four inches of arch depth.

CHIMNEYS

The chimney construction specifications given here are adapted from the national building code for low-heat appliances. Warm air furnaces and steam boilers operating at not over 50 pounds per square inch gauge-pressure are classified as low-heat appliances. These include the heating

262

units for most residential and apartment buildings, as well as many commercial buildings.

Height: Chimneys for low-heat appliances should extend at least three feet above the highest point where they pass through a flat or low-pitched roof of a building and at least two feet higher than any ridge or high portion of the building within 10 feet (Fig. 19-6).

Support: Masonry chimneys should be supported on foundations of masonry or reinforced concrete or other non-combustible material having a fire resistance rating of not less than three hours.

Corbeling: No chimney should be *corbeled* from a wall (supported by a structure projecting from the wall) more than six inches; nor should a chimney be corbeled from a wall which is less than 12 inches in thickness unless it projects equally on each side of the wall. In the second story of two-story dwellings, corbeling of chimneys on the exterior of the enclosing walls may equal the wall thickness. Corbeling should not exceed one-inch projection for each course of brick projected.

Changes: No change in the size or shape of a chimney, where the chimney passes through the roof, should be made within a distance of six inches above or below the roof joints or rafters.

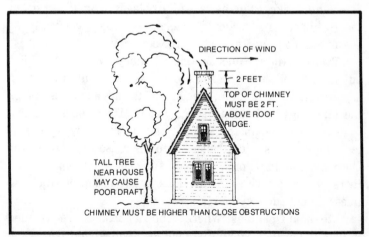

Fig. 19-6. Chimneys must extend at least two feet beyond the highest point of the building. If a tall tree is close to the house, poor drafts may result.

Materials: Masonry chimneys for low-heat appliances should be constructed of solid masonry units or of reinforced concrete. Chimneys in dwellings, chimneys for domestic-type low-heat appliances, and chimneys for building heating equipment for heating a total volume of occupied space not to exceed 25,000 cubic feet must have walls not less than four inches thick. In other buildings and for other low-heat appliances, the thickness of chimney walls should be not less than eight inches. Rubble stone masonry should not be less than 12 inches thick.

Liners: Masonry chimneys for low-heat appliances should be lined with approved fireclay liners not less than 5/8 inch thick, or with other approved liner of material that will resist heat without softening or cracking at a temperature of 180° Fahrenheit. Fireclay flue liners should be installed ahead of construction of the chimney, as it is carried up, and should be carefully bedded in type M, type N, or fireclay mortar with close-fitting joints left smooth on the inside.

Chimney and fireplace construction is shown in Fig. 19-7. In masonry chimneys with walls less than eight inches thick, liners are built separately from the chimney wall and the space between the liner and masonry is not filled; only enough mortar is used to make a good joint and hold the liners in position. Flue liners start from a point at least eight inches below the intake, or in the case of fireplaces, from the throat of the fireplace. They should extend, as nearly vertically as possible, for the entire height of the chimney.

Two or More Flues in One Chimney: Where two flues adjoin each other in the same chimney, with only flue lining separation between them, the joints of the adjacent flue linings must be staggered at least seven inches. Where more than two flues are located in the same chimney, masonry wythes at least four inches wide, bonded into the masonry walls of the chimney, should be built between adjacent flue linings so that there are not more than two flues in any group of adjoining flues without such wythe separation.

Cleanout Openings: Where cleanout openings are provided in chimneys, they should be equipped with metal doors and frames arranged to remain tightly closed when not in use.

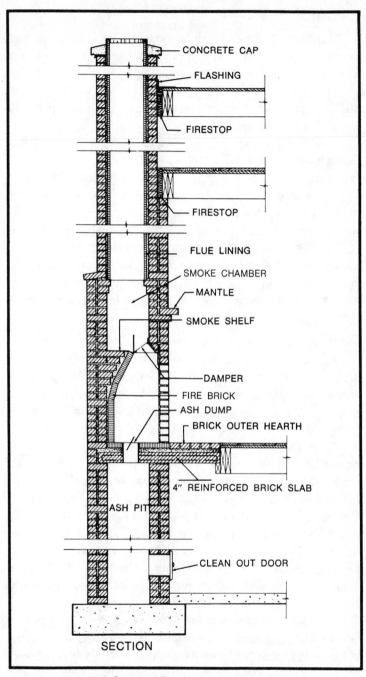

CONCRETE CAP

FLASHING

FIRESTOP

FIRESTOP

FLUE LINING

SMOKE CHAMBER

MANTLE

SMOKE SHELF

DAMPER

FIRE BRICK

ASH DUMP

BRICK OUTER HEARTH

4" REINFORCED BRICK SLAB

ASH PIT

CLEAN OUT DOOR

SECTION

Fig. 19-7. Standard fireplace and chimney design.

Connections

Standard flue lining sections are available with openings to receive smoke pipe connections; however, if the opening is to be cut, it should be done before the lining is installed.

Flue lining can be cut by filling the section with damp sand, tamped solid, and tapping with a sharp chisel with a light hammer along the line where the cut is desired.

The smoke pipe connection should enter the side of the flue through a thimble or flue ring. Such rings are available with inside diameters of 6, 7, 8, 10, and 12 inches, and lengths of 4 1/2, 6, 9, and 12 inches. These rings should be built in as the work progresses and must be made airtight at all points. The metal smoke pipe should not enter farther than the inner face of the flue, and the joint around the pipe should be made airtight with boiler putty or asbestos cement. The top of the smoke pipe should be not less than 18 inches below the ceiling, and no wood or combustible material should be placed within 6 inches of the thimble.

Chimney Tops and Flashing

The tops of chimneys above the roof offer unlimited architectural possibilities. In the more ornamental types, round flue linings may be more advantageous and in this event, the same round lining should be used for the full height of the flue.

Regardless of the architectural design, certain structural details should always be followed. The flue lining should project four inches above the chimney cap which should be finished with a straight or concave slope to direct the air current upward at the top of the flue and also to drain water from the top of the chimney. It is preferable to form the cap with sufficient projection to serve as a drip and keep the walls dry and clean. The masonry flue walls should be bonded, either by carefully staggering the joints or by using non-corrosive metal ties.

Fireclay chimney tops or pots are available for use on new or old chimneys. Care should be exercised in selecting these units so that the effective area of the flue is not reduced. They are made to fit over the flue linings, one to each flue, and should be set in type M mortar finished to form a wash.

At the intersection of chimney and roof, the connection should be made weathertight by means of flashing and counter flashing, preferably of copper or other rust-resisting metal, arranged to allow for any vertical or lateral movement between the chimney and roof.

Chimney Test and Repair

After the chimney has been completed and the mortar thoroughly hardened, and before any appliances are connected, a *smoke test* should be made on each flue. This is done by building a *smudge fire* at the bottom of the flue and, while the smoke is flowing freely from the flue, covering the top tightly. Escape of smoke into other flues or through the chimney walls indicates openings that must be made tight before the chimney is used.

Repairing of leaks is usually difficult and expensive, and it is, therefore, much more satisfactory and economical to see that construction is properly executed as the work progresses.

Smoke passages and chimney flues should be kept clean. An accumulation of soot may cause a chimney fire with the consequent danger of sparks igniting the roof or causing damage to the chimney itself which would permit passage of fire through the fire walls.

The most efficient method of cleaning a chimney flue is to use a weighted brush or bundle of rags lowered and raised from the top.

The tops of unlined chimneys may have to be rebuilt every few years due to the disintegrating effect of smoke and gases on the mortar. The chimney should be taken down to a point where the mortar joints are solid. The new top should be built with fireclay flue lining of the same size as the old flue.

FIREPLACES

There are several types of residential fireplaces. Within each of these general classifications, there are many individual designs, but the operation and construction remains the same.

Single-Face Fireplace: This is the familiar one-opening design. Its origin is obscure, but its design has become a

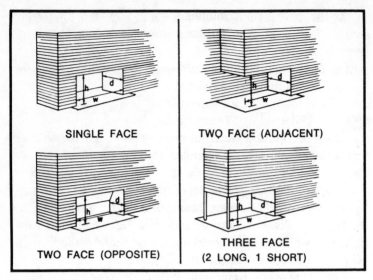

SINGLE FACE

TWO FACE (ADJACENT)

TWO FACE (OPPOSITE)

THREE FACE
(2 LONG, 1 SHORT)

Fig. 19-8. Basic fireplace types (Courtesy BIA).

science, and a great deal of information as to proper opening, damper, and flue sizes is available. This means there is no longer any element of mystery or luck connected with its successful construction.

Multi-Face Fireplace: Although most generally associated with contemporary design, this type of fireplace is also of ancient origin. For example, the so-called *corner fireplace*, which has two adjacent sides open, has been in use for several hundred years in the Scandinavian countries (Fig. 19-8).

Another modification of this type of fireplace is one in which the two opposite faces are open. This is popular as a free-standing room divider, since it has the advantage of providing fire and warmth for two rooms.

Other forms of this design include those with three or even four sides open. Such designs, shown schematically in Fig. 19-8, are often impressive and highly effective when properly used. However, some present certain design problems which must be solved before satisfactory performance may be expected. Adequate draft must be obtained through the use of oversize flues and controlled face opening sizes. The possibility of cross-drafts through the fireplace, which could at least momentarily carry smoke out into the room, must be considered. If these possibilities seem likely, some provision, such as glass fire screens on one or more of the face openings, should be used.

Careful consideration should be given to the size of the fireplace best suited to the room in which it is located, not only from the viewpoint of its appearance but of its operation as well. If it is too small, it may function properly, but will not produce a sufficient amount of heat. If too large, a fire that would fill the combustion chamber would be entirely too hot for the room.

A room with 300 square feet of floor area is well served by a fireplace with an opening 30 inches to 36 inches wide. For larger rooms, the width may be increased.

Fireplace openings should not be too high; for the usual width of opening, the height above the hearth is seldom more than 32 inches. Other dimensions are given in Table 19-2.

Dimensions may be varied slightly to meet regular brick courses and joints, especially where the facing or trim is of brick.

Combustion Chamber

The shape and depth of the fireplace's combustion chamber influence both the draft and the amount of heat radiated into the room. Recommended depths are given in Table 19-2. Again, these dimensions might be varied slightly.

The slope of the back throws the flame forward and leads the gases with increasing velocity through the *throat*. The

Table 19-2. Recommended Fireplaces and Flue Dimensions

Type	Height of Opening	Size of Hearth	Flue Size (Modular)
Single Face	29″	16″ deep, 30″ to 36″ wide	12″ × 12″
		16″ deep, 40″ wide	12″ × 16″
	32″	18″ deep, 48″ wide	16″ × 16″
Double Face (Adjacent)	26″	16″ deep, 32″ wide	12″ × 16″
	29″	16″ to 20″ deep 40″ to 48″ wide	16″ × 16″
Double Face (Opposite)	29″	28″ deep 32″ wide	16″ × 16″
	27″	28″ deep, 36″ to 40″ wide	16″ × 20″
Three Face (2 long, 1 short)		32″ to 40″ deep, 36″ to 44″ wide	20″ × 20″

Fig. 19-9. Firebrick is used for interior surface of fireplace.

slope, along with the splay of the sides, also gives the maximum radiation of heat into the room.

The combustion chamber should be lined with firebrick laid with thin joints of fireclay mortar (Fig. 19-9). The back and end walls of the average size fireplace should be at least eight inches thick, and thicker for larger sizes, to support the chimney load above.

Throat

Because of its effect on the draft, the throat of the fireplace should be carefully designed (Fig. 19-10). It should be not less than six inches and preferably eight inches above the highest point of the fireplace opening. The sloping back extends to the same height and forms the support for the back of the *damper*.

A metal damper should be placed in the throat, and should extend the full width of the fireplace opening. The design should include a plate which opens upward toward the back. Such a plate, when open, forms a barrier to stop down draft and deflect it upward into the ascending column of smoke. When the fireplace is not in use, the damper should be kept closed to stop drafts and heat loss from the room, as well as to keep out dirt from the flue.

Metal dampers are available in several types. Some form the damper throat and supporting lintel in one piece over the fireplace opening. This type has the advantage of producing a smooth throat passage and simplifying the masonry work.

Smoke Shelf and Chamber

The location of the throat establishes the position of the *smoke shelf*. This shelf should be directly under the bottom of the flue, extending the full width of the throat. It must be constructed horizontally, and not sloped, because its purpose is to stop the down draft.

The space above the shelf is the *smoke chamber*. The back wall of the chamber is built straight, the side walls are sloped uniformly to the center to meet the bottom of the flue lining, and the front wall above the throat is also sloped to the flue lining.

Metal lining plates are available for the smoke chamber and are effective in giving the chamber its proper form and smooth surfaces. They also simplify the bricklaying.

Fireplace Flue

Relatively high air velocities through the throat and flue are desirable. The velocity is affected by both the area of the

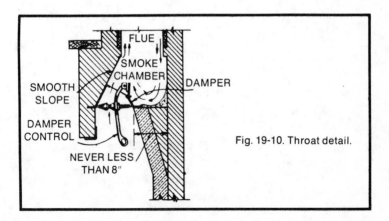

Fig. 19-10. Throat detail.

flue and the height of the chimney. The proper sizes of fireplace flues are given in Table 19-2.

The fireplace should have an independent flue entirely free from other openings or connections, and the first section of flue lining must start at the center line of the fireplace opening. This is important in obtaining a positive and uniform draft over the full width of the fireplace. The flue lining should be supported on at least three sides by a ledge of projecting brick, finishing flush with the inside of the lining.

Construction Details

There are two methods of building a fireplace and chimney assembly. One is to complete the fireplace and chimney in one operation as the work progresses. The facing may be omitted until later, in which case suitable ties are built in for properly bonding the facing in place.

The other method is to form a rough opening for the later construction of the fireplace. In this case, the end and back walls are built eight inches thick; a steel lintel is set below the shelf level to carry the front wall above; the back and sides of the smoke chamber are formed, and the flue is built in. Care must be taken in building the fireplace to finish the work tightly to the underside of the previously built chimney. This will prevent leaks and the possibility of fire hazard.

Part 4 : Stone

20

Structural
Clay Tile Masonry

Hollow units made of burned clay or shale are called *structural clay tile*. Several common types of clay tile and their dimensions are shown in Figs. 20-1, 20-2, and 20-3. Structural clay can be used for both exterior and interior walls, depending on the type. It can also be used for partitions, and for brick backer.

The masonry units are made by forcing a plastic clay through special dies. The tiles are then cut to size and burned in the same manner as brick. The amount of burning depends upon the grade of tile being manufactured.

The hollow spaces in the tile are called *cells*. The outside wall of the tile is called the *shell*, and the partitions that divide the tile into cells are called *webs*. The shell should be at least 3/4 inch thick and the web should be 1/2 inch thick.

CLASSIFICATIONS

In side-construction tile the cells are horizontal, while in end-construction tile the cells are placed vertically. There is no particular reason to prefer one type over the other. Both side-construction and end-construction tile are made in the following structural clay classifications:

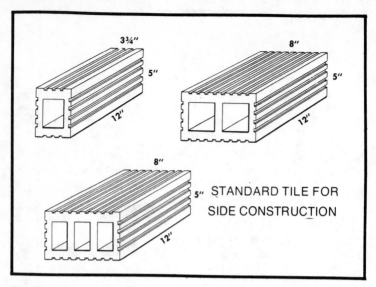

Fig. 20-1. The most commonly used types of structural clay tiles and their dimensions. Above are standard tiles for side construction.

Structural Clay Load-Bearing Wall Tile—Tile in this classification may be divided into two types depending upon use:

(1) Tile used for the construction of exposed or faced load-bearing walls. This tile is designed to carry the entire load including the facing material. The facing may consist of stucco, plaster, or other suitable material.

(2) Tile used for backing up brick walls that bear loads and those that do not. The facing or outer tier of brick in these walls is bonded to the backing tile by headers. The wall load is supported by both the facing and the backing. The inside face of this tile is scored, making it possible to plaster it without using *laths* (thin strips of wood used as ground work, for plaster).

ASTM Classification— The American Society for Testing and Materials gives two grades of structural load-bearing tile based on resistance to weathering.

Grade LB. This grade is suitable for general use on masonry construction where not exposed to weath-

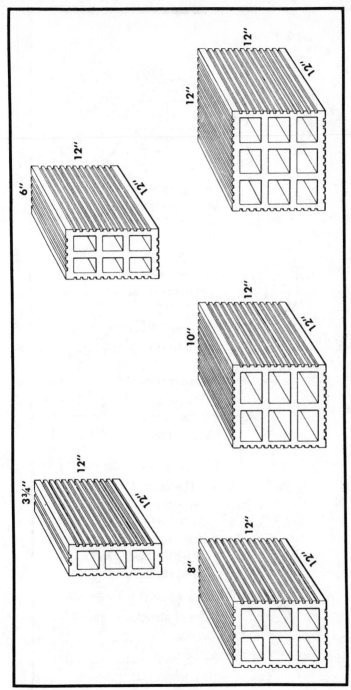

Fig. 20-2. Standard tiles for end construction.

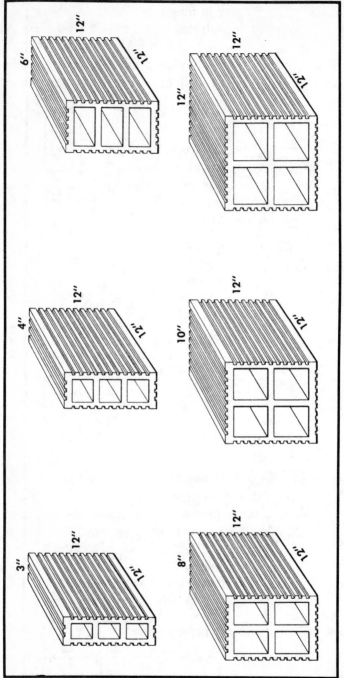

Fig. 20-3. Standard tiles for partitions.

ering or for use in masonry exposed to weathering, provided it is protected with at least three inches of facing.

Grade LBX,. This tile may be used in masonry exposed to weathering with no facing material required.

Structural Clay Non-Load-Bearing Tile— This classification includes:

Partition tile used in the construction of non-load-bearing partitions or for backing non-load-bearing brick walls.

Tile used for *lining* the inside of a wall in order to provide a surface that may be plastered. These tiles also provide an air space between the plaster and the wall.

Tile used to *fireproof* steel columns and beams.

Structural Clay Facing Tile— There are two classes of structural clay facing tile based on shell thickness: *standard* and *special duty*. The surface finish of this tile closely resembles the surface finish of face brick (Fig. 20-4). There are two types in each class.

Type FTX. This tile is better in appearance and the easier of the two types to clean.

Type FTS. This tile is inferior in quality to type FTX but is suitable for general use where some defects in surface finish are not objectionable.

Structural Glazed Facing Tile— The exposed surface of glazed tile has either a ceramic glaze, a salt glaze, or a clay coating. These tiles are used where stainproof, easily-cleaned surfaces are desired. They can be obtained in many colors and they produce a durable wall with a pleasing appearance.

Special Units— In addition to standard stretcher units, special units are available for use at window and door openings and at corners. Check with the dealer for these units.

PHYSICAL CHARACTERISTICS

Structural clay tile is characterized by several physical attributes:

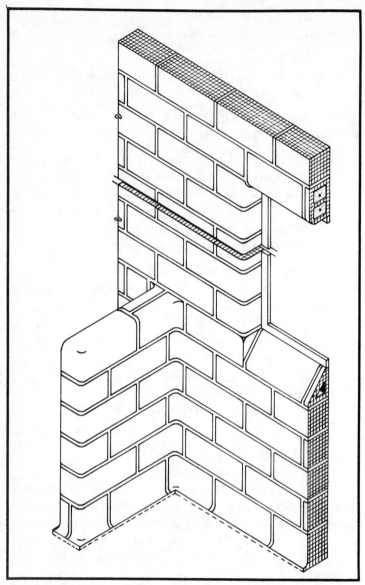

Fig. 20-4. A wall built of structural clay facing tile (Courtesy BIA).

Strength

The compressive strength of the individual tile depends upon the materials used, upon the method of manufacture, and upon the thickness of the shells and webs. A minimum compressive strength of tile masonry of 300 pounds per square inch based on the cross section may be expected. The tensile strength of structural clay tile masonry is small. In most cases, it is less than 10 percent of the compressive strength.

Abrasion Resistance

As with brick, the abrasion resistance of clay tile depends primarily upon its compressive strength. The stronger the tile, the greater its resistance to wearing. The abrasion resistance decreases as the amount of water absorbed increases.

Weather Resistance

Structural clay facing tile has excellent resistance to weathering. Freezing and thawing action produces no deterioration. Only portland-cement-lime mortar or masonry cement should be used if the masonry is exposed to the weather.

Heat- and Sound-Insulating Properties

Walls containing structural clay tile have better heat-insulating qualities than do walls composed of solid units, due to dead air space that exists in tile walls. Clay tile's resistance to sound penetration is good but is still somewhat less effective than other types of masonry.

Fire Resistance

The fire resistance of tile walls is considerably less than the fire resistance of solid masonry walls. The fire resistance can be improved by applying a coat of plaster to the surface of the wall. Partition walls of structural clay tile six inches thick will resist a fire of 1700° Fahrenheit for one hour.

Weight

The solid material in structural clay tile weighs about 125 pounds per cubic foot. Since the tile contains hollow cells of various sizes, the weight of tile varies, depending upon the manufacture and type. A six-inch tile wall weighs approximately 30 pounds per square foot, while a 12-inch tile wall weighs approximately 45 pounds per square foot.

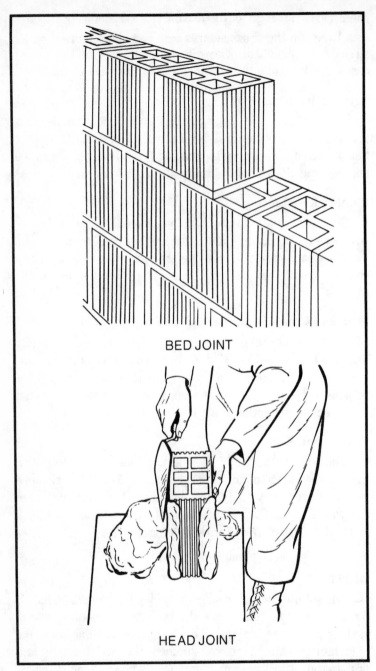

BED JOINT

HEAD JOINT

Fig. 20-5. How to build a wall of end-construction clay tile, with bed joints and head joints.

282

WORKING WITH STRUCTURAL CLAY TILE

In general, the procedures for using structural clay tile are the same as for using brick. The same tools are used here also.

When using end-construction units, the bed joint is made by spreading a one-inch thickness of mortar on the shell of the bed tile but not on the webs, as shown in the bottom of Fig. 20-5. The mortar should be spread for a distance of about three feet ahead of the laying of the tile. The position of the tile above does not coincide with the position of the tile below since the head joints are to be staggered. The web of the tile above will not contact the web of the tile below and any mortar placed on the web is useless.

The *head joint* is formed by spreading plenty of mortar along each edge of the tile, as shown in Fig. 20-5, and then pushing the tile into the mortar bed to its proper position. Enough mortar should be used to cause the excess mortar to squeeze out of the joints. This excess mortar is cut off with a trowel. The head joint need not be as full as recommended for head joints in brick masonry unless the joint is to be exposed to the weather. Clay tile units are heavy, making it necessary to use both hands when placing the tile in position in the wall. The mortar joint should be about 1/2 inch thick, depending upon the type of construction. *Closure joints* are the same as in brick masonry.

Side-construction bed joints are made by spreading the mortar to a thickness of about one inch about three feet ahead of the work. The furrow required for brick work is unnecessary for clay tile.

For head joints, as much mortar as will adhere is spread on both edges of tile. The tile is then pushed into the mortar bed against the tile already in place until it is in its proper position. Excess mortar is cut off. Mortar joints should be about a half inch.

Using Structural Clay as Brick Backup

The example in Fig. 20-6 shows side-construction backing tile, four inches wide, five inches high, and 12 inches long. The height of one tile is equal to the height of two common

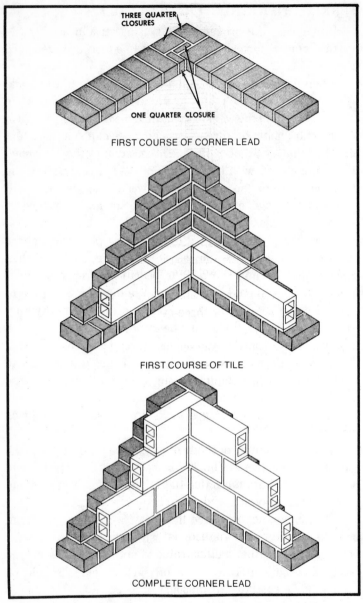

THREE QUARTER CLOSURES

ONE QUARTER CLOSURE

FIRST COURSE OF CORNER LEAD

FIRST COURSE OF TILE

COMPLETE CORNER LEAD

Fig. 20-6. A brick wall with tile backup (side-construction tile).

non-modular bricks with a half-inch mortar joint. Bed joints on
the tile are as thick as required to make the tile level with
every other brick course (usually this comes to one-half inch

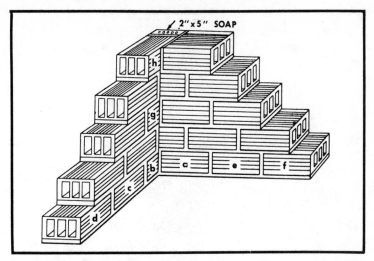

Fig. 20-7. Building an all-tile wall using eight inch thick, side-construction units. "Soap" avoids exposure of open cells to the elements.

also). The bottom course is all brick laid in headers. Note the closure at the corners of three-quarter and one-quarter split bricks.

After several brick courses are laid up, the tile is laid on top of the header course as shown. The tile is laid as described above, and the wall is built up from there to the desired height and length.

All-Tile Construction

When a wall is to be made of all tile, eight-inch or thicker tile is used. The tile in Fig. 20-7 is also five inches high and 12 inches long. Side-construction tile is used here, except for the corner tiles, which are end-construction (b and g in Fig. 20-7) in order to avoid exposure of the open cells at the face of the wall. An alternate method is to use a *soap* or thin, end-construction tile, as illustrated at the top of the corner in Fig. 20-7.

As is the case for all masonry construction, it is always best to make a dry run before beginning to lay clay tile. Each tile should be individually leveled and plumbed as discussed for concrete block. After the corners are built up, the tiles are laid as described above. Use corner pins and lines as you would for any other type of masonry construction.

21

Building
With Stone—
Tools and Materials

Stone is by far man's oldest building material. Wherever ancient civilizations are found, their edifices are found to have been constructed of stone. Archeologists and others are often astounded, in fact, by the remarkable walls, temples, and other structures made by so-called primitive people using only crude hand tools.

The pyramids of Egypt are considered one of the wonders of the world. Built almost entirely of huge stones weighing many tons, the pyramids and their builders are the subject of much conjecture. How, for example, did the Egyptians raise these monstrous building blocks on top of one another? There are many theories, but no one knows for sure, and it remains a most intriguing mystery. The pyramids and other ancient ruins also stand as mute evidence of the durability and workability of stone.

Another of the marvelous qualities of stonework is also exemplified by the ancient buildings. There was, in ancient times, only a crude form of mortar, if any. How do these buildings survive without something holding the stones together? The answer is that stone is the only building material which does not need mortar, nails, or any form of attachment to hold it together. The abrasive quality and great weight of the individual stones makes them difficult or

Fig. 21-1. Building stone can be used in a number of ways to enhance the landscape of a home (Courtesy Building Stone Institute).

impossible to move. The elements will eventually destroy any building, even a stone one, and mortar will in time decay, but stone stays put through wind, hail, and rain—even while erosion takes place. Someday, the Great Wall of China will be reduced to sand (unless it is restored, of course), but it will probably survive at least another thousand years.

MODERN STONE

You, of course, are probably less concerned with stone lasting an eon than you are with construction that will survive your own lifetime. Actually, all of the building materials in this book should last a lifetime, but a stone structure that is built correctly will last your lifetime and many more after that.

Stone is not only durable, but it is very attractive, too (Fig. 21-1). While brick may be considered by some to be more elegant, and concrete more workable, there are varieties of stone for every occasion, including indoor uses that can be very beautiful (Fig 21-2). A properly designed stone wall will be both elegant and practical, and stone is the best material for attaining the rustic look so popular today (used brick, perhaps, may be a close second). Remember that the chief ingredient in concrete is stone aggregate, which is what gives it strength and durability.

There are several hundred varieties of stone, classified by mineral content, color, texture, and the region in which it is quarried. The most popular types of natural stone are limestone, sandstone, marble, slate, quartzite, and granite.

Fig. 21-2. Don't overlook stone for interior use. It is especially attractive for fireplaces, planters, and room dividers (Courtesy BSI).

Each type is further subdivided. Granite, for example, comes in Oxford gray, French Creek black, blue diamond, and many others. There is Vermont marble, Alabama marble, Missouri marble, Tennessee marble, and Colorado marble. Sandstone is sold as New York bluestone, Amherst sandstone, golden sandstone, or brownstone. You can buy Allegheny quartzite, Tennessee variegated quartzite, or crab orchard quartzite.

No one particular stone is best for any single project. Stone is bought primarily on the basis of local availability. An architect or decorator may be particularly enamored of bluestone (especially if he lives in New York). But it would be extremely costly to build a house in Colorado of New York bluestone. The same applies to a home in New York built of Colorado marble.

The only way to decide on what stone is best for your project is to visit local stone dealers and see what they have to offer. If only one type will do, a dealer can probably get it for you, but you'll pay dearly for your fixation. Stone is heavy stuff, and transportation costs are frightening. There will no doubt be many types and colors available, and each is as good as the other for almost every project (with the exception of *rubble* stone which is discussed below). The color should blend in with your home and the landscape, and the stone should be reasonably easy to work with. Those are your prime considerations.

It may be that you won't have to visit the stone dealer at all. In certain parts of the country, natural stone may be right in your own back yard. These stones ordinarily appear in the form of boulders (called *rubble* in the trade). Boulders make fine building material, but they are not right for every project, because they are easier to topple and harder to cut.

Rubble can be used for an informal garden wall, a rustic fireplace, or a retaining wall. Building techniques for these projects differ, however, from heavier construction work that requires the longer, flatter and softer *ashlar* stone. Ashlar stone is usually quarried and often cut into regular designs. You can buy ashlar stone as it comes from the quarry, or cut into various rough or exact thicknesses. Stone purchased ''as is" is much cheaper than stone that has been factory-cut into neat rectangles and squares.

The one great difficulty with stone lies in the cutting. Stone can be cut, but it is a rather tedious and back-breaking job, so weigh this against the added cost of paying for precut stone and make up your own mind as to which you should buy. You should also consider the type of project. A formal patio would be difficult to build with uncut stone; but you don't need neat, exact pieces for a retaining wall.

Rubble stone cannot be cut at all, not if you're looking for a certain size and shape. The only time you may want to use rubble other than as it comes from nature is as a rough veneer for a house front or a fireplace. You can break rubble by hitting it with a sledge hammer or banging it against a hard surface. This exposes the crystalline inner surface, which can be very attractive in some stones. Ordinarily, rubble is simply piled up with or without mortar.

SELECTING STONE

In addition to the qualities listed above, you should consider a few other factors in your planning. You should decide whether your project will be laid up wet (with mortar) or dry (without mortar). More formal installations such as fireplaces and barbecues are usually put up wet. Retaining walls, planters, and low walls are usually laid dry. But there are no hard and fast rules for this. Stonework is the least orthodox and most creative of all the building arts. Here, you can do your thing without having to worry about whether or not it looks professional.

Stone can also be used as veneer. Block, wood, or some other material can be used structurally, and the stone can be laid with its flat side against the backup. If you prefer this method, the stone must be cut to an even thickness—four inches for most stone, or two inches for marble. Mortar is used here, and the stone is attached to the backup by means of clips (this is discussed below).

For a small project, you may want to select individual pieces of stone and pay by the piece or by the square foot. Most jobs will require larger amounts, though, in which case stone

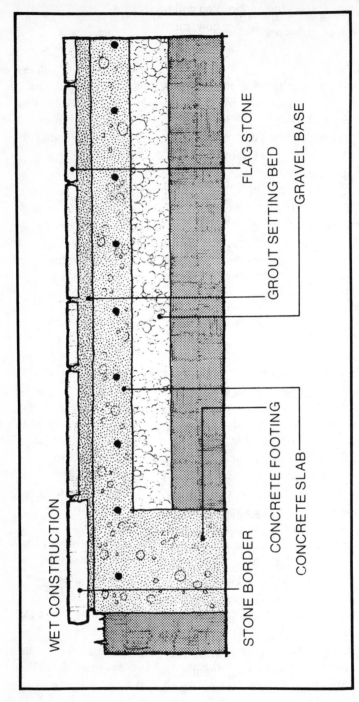

WET CONSTRUCTION

STONE BORDER

CONCRETE FOOTING

CONCRETE SLAB

FLAG STONE

GROUT SETTING BED

GRAVEL BASE

Fig. 21-3. Patio construction, with mortar.

may be bought by the ton. You will need personal help from the stone dealer in deciding how much stone to buy, since the variations are great. The range may vary from 35 to 50 square feet of frontage per ton. The retaining walls illustrated in this chapter took three tons of stone. Steps require several large, extra pieces.

Precut stone is usually sold in multiples of six inches. Sizes start at nominal 12 × 12 inches and go up to 24 × 36. Larger sizes may be purchased, but are often proportionately more expensive. Smaller ones, probably cut from waste or damaged pieces, may also be available. Cut stone are usually used with mortar, although it's not necessary. For this reason, the dimensions of cut stone, as noted above, are nominal, allowing space for mortar joints. The actual size is usually 1/2 inch less all around.

The thickness of cut stone varies. Since this type of stone is usually used flat, in patios or walks, it makes a difference whether you lay it dry or wet (see Figs. 21-3 and 21-4). If you plan on using mortar, the smaller thicknesses are fine. For dry construction, the pieces should be at least one inch thick to prevent cracking.

When buying stone, remember that only one face will be seen. Flagstone need be smooth only on one side. The bottom side may be rough and the thickness may vary (although varying thicknesses make it harder to install). The same applies to stone used for veneer, except that the thickness should be reasonably uniform. If you want to lay the stone in a formal, geometric pattern, the edges must be square, but not necessarily smooth. Random shapes should be smooth on one side also. These random-shape stones may be difficult to lay, but they are cheaper and have a more rustic look. The various types of patterns are illustrated in Fig. 21-5.

When buying ashlar wallstone, remember it is the front edge only that counts. For an informal wall, a smooth, straight edge is not necessary. Construction is easier if one edge is perfectly straight, but slight variations and curved edges are perfectly acceptable, as long as the overall lines are pleasing. Some individual pieces may be very rough and jagged, but the overall effect is a long, gentle, neat curve. The rules for stonework are very flexible.

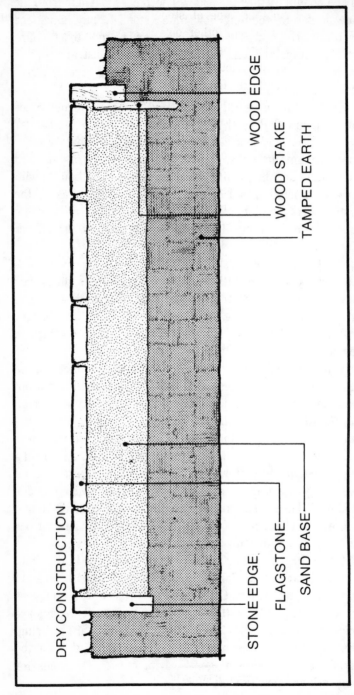

Fig. 21-4. Patio construction, without mortar.

WOOD EDGE

WOOD STAKE

TAMPED EARTH

DRY CONSTRUCTION

STONE EDGE

FLAGSTONE

SAND BASE

There is another type of stone that you may wish to consider for gardens or other outdoor areas not subject to hard use: aggregate (pebbles or chips). These decorative stones come in a variety of colors. You can buy them in 100-pound bags or by the ton.

Another possibility is artificial or manmade stone, usually made of plastic, sometimes with ground-up stone to give it a more authentic look. These stones are cheaper and easier to use for veneer. They can't be used for true construction, however. *Patio blocks* are somewhat similar to manmade stone, and are made very much like cinder block. These are lightweight and are an acceptable alternative for horizontal applications, but not for installations where strength is required. In general, natural stone is much stronger, more attractive, and a much better all-around product than manufactured pieces. Where cost is a vital factor, however, you may wish to consider substitutes.

OTHER MATERIALS

If your project is going up dry, you may need no other materials at all. Where soil is naturally sandy and not overly

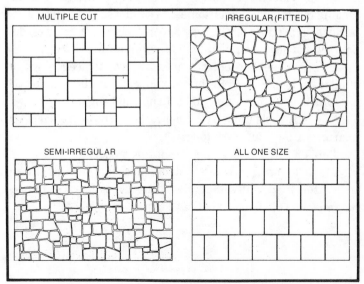

Fig. 21-5. Patterns for laying flagstone (Courtesy BSI).

organic, stone can be laid right in the ground. Most soils, however, even in dry construction, should be dug out and replaced by four inches of sand or stone dust.

Sand—The sand doesn't have to be the high-quality sand used for mortar. Beach or bank-run sand is okay as long as it doesn't contain a lot of organic material.

Stone Dust—This is actually a little more permanent than sand for use as a base. It costs a little more, but may be worth it. Check local prices. Usual cost is about 25 percent more than sand.

Mortar—Mortar mixtures are simpler for stonework than they are for brick and concrete block. The usual proportion is one part portland cement to two parts of sand, mixed with water to a mud-like consistency. The sand here, of course, must be the same fine type used for all mortars. Don't use the same sand bought for the base, as described above.

Metal Ties—The ties used in stonework veneers are similar to those used for brick and stone.

TOOLS

For most stonework, the biggest requirements are that old strong back and a good pair of hands. You will need a spade for digging out the earth and some sort of measuring and leveling devices, depending on the type of construction. If you intend to do dry construction you will need no further tools. For mortared veneer or other wet work, you'll need a trowel similar to the one used for brick or block construction, and the other tools used for mixing, measuring, and aligning.

The only tools unique to stonework are those needed if you plan on cutting your own stone:

Cold chisel: The best tool for cutting stone.

Mash hammer: A small sledge-like hammer for hitting the chisel. Also called a drilling hammer. A heavy ball-peen hammer will do if you do not have a mash hammer.

22

Laying Rubble Stone

Erecting rubble stonework can be as easy or as hard as you want to make it. The early colonists were not skilled stone masons, but they put up many low stone walls as property boundaries which are still around today. As pointed out in Robert Frost's poem, *Mending Wall*, such walls would have to be fixed up once in awhile. The field stones, which were simply placed on top of each other as they were removed from the farmland, were subject to heaving and displacement because of the hard New England frosts.

The type of wall Frost was talking about is known in the trade as *random rubble*. As constructed by the early colonists, the wall was made by simply placing whatever boulders were at hand on top of each other (Fig. 22-1). For a rustic look, and a wall that isn't designed to do much structurally, this is perfectly acceptable construction.

If you want your rubble wall to last, however, you should choose your stones carefully. If rounded stones are all that is available, they are all right, but the best ones for wall-building are those that have flat sides. The ideal building stones have six flat sides, but you'll find few of those. If they are reasonably flat at the top and bottom, so much the better. The only kind of stones you shouldn't use are odd-shaped ones, which are not only rather ugly, but hard to build with.

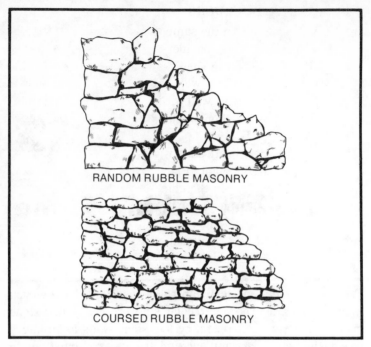

RANDOM RUBBLE MASONRY

COURSED RUBBLE MASONRY

Fig. 22-1. The difference between "random" and "coursed" rubble.

Look for colorful, good-looking stones, and ones that you can handle without too much trouble. The larger the stone, the more secure your wall will be; but giant boulders aren't going to be of much use unless you have a bulldozer to move them. If possible, get a good, strong helper, and choose only those stones which the two of you can lift.

Don't overlook smaller stones, though, you'll need some of them to fill chinks, and they will be easier to lift as the wall gets higher. The best walls have larger stones at the bottom and smaller ones near the top. This is not only easier to build, but it looks better, too.

You will notice that stones have *grain* to them. The geological strata run in a certain direction. Stones should be laid with the grain horizontally, as it occurs naturally. Avoid stones that are high vertically. They can be used here and there, but always lay them with the grain sideways, and don't try to lay them on their side. The best stones for building are long and somewhat flat.

WORKING WITH BOULDERS

Not everyone means the same thing when they speak of a "boulder." As used here, boulder means natural stones that are more or less round in shape. Although this type of stone doesn't make a very sturdy wall, it can be used if you build a pyramid-type wall. Such a wall is wide at the bottom and small at the top. A cross section would have a roughly triangular shape.

To lay this type of wall, start with a layer of stone about as wide as the desired height. Place your next layer on top, centering each boulder in the chinks between the one below. Simply lay your rows one on top of the other until the last row consists of single stone. This type of wall is the most primitive of all, but it has a nice, old-fashioned look that many people find pleasing. It is especially attractive on large acreage or farms.

You can make a boulder wall much sturdier by using mortar, although this detracts from its chief virtue of rusticity. If you use mortar, be sure to give the wall a good footing to prevent heaving and cracking. A standard concrete footing, as described in previous chapters, is best but it may be impractical. The usual method is to dig down a foot or so and use some stone below ground as your footing. Make the footing 1 1/2 to two times as wide as the wall and fill in chinks with smaller stones and mortar. Use masonry cement (with lime) to mortar any stone wall.

USING FLATTER STONES

No rubble stone is going to be perfectly flat. But if you can find stone with two fairly straight sides, you'll be able to build a more solid wall. Flat or rectangular shaped stones can be used in a *coursed* stone wall. The stones are laid in courses (admittedly not too even) and usually have sides that are straight or slightly tapered toward the top.

With this type of construction, you must be more selective in your stone selection. In areas where frost is a hazard, start below grade and select the biggest stones to use as a footing. Since a coursed wall is more formal, you should lay it out carefully, using batter boards and line to make sure that the wall is straight and plumb. Drive in stakes every eight feet or so to keep the line taut.

Pick out large stones that are approximately even in height and use them for your first course. You can use uneven stones, but smaller stones should fill in to make each bed joint more or less continuous. Don't lose a lot of sleep over straight lines, however, since this type of construction is an inexact science, at best.

Your wall may be perfectly plumb if you wish, but it is a little easier to batter the wall slightly as you approach the top. Use smaller and smaller stones as you go up. Each wall should be at least two large stones wide, with the longest side laid lengthwise. A bonding stone should be laid between the two sides approximately every eight square feet. Corners should be put up first, at least in the beginning.

It is important in this and all types of stonework not to worry about even courses and exact plumb. Stone just doesn't lend itself to neat, fine lines. The only rule to follow is to keep the head joints from following the same vertical line. Each stone should be laid so that the center is as close as possible to the joint below, avoiding *runs* in head joints with continuous vertical lines.

There is no particular difference in basic construction when the wall is to be made with mortar. Mortar is placed on the bed joint and the stone is buttered very much as in bricklaying and concrete block. Since this is a very erratic art, you will find that you are filling in with mortar a lot, instead of having a neat mortar joint. It is better to make a neat joint than to have a lot of excess mortar oozing out and staining the stone.

You should also use smaller stones to fill in chinks between the larger ones. Leave as little open space as possible, whether working with mortar or not.

CUTTING RUBBLE STONE

As discussed in the previous chapter, there is no good way to cut rubble stone. The best approach is to use it as is or forget it. But if you feel you *want* to cut it for use as veneer or for nice squared corners in a fireplace, there *is* a way.

For veneer, it is really best to use ashlar stone (see the next chapter). Otherwise, just go at it with a sledge hammer and try to get a fairly even break. If it's square corners you are after, you can accomplish this the hard way by chipping a line all around with a cold chisel, then whacking the chisel with a mash hammer. Chances are you'll ruin more than you cut. If you have a lot of square cutting to do, you may want to invest in a masonry-cutting blade for your circular saw.

BUILDING RETAINING WALLS

A rubble retaining wall is basically the same as a regular wall (Fig. 22-2). The differences have to do with the fact that a retaining wall is subject to great pressure from one side. The pressure does not come so much from the earth itself—although, of course, it is a factor—as from water which builds up behind the wall.

Boulders are not a very good choice here, but can be used if the wall is not too high. Actually, there is no difference at all

Fig. 22-2. Boulders can be used to form retaining walls simply by setting them in the side of the hill (this is for small walls only).

Fig. 22-3. A random rubble retaining wall built with mortar.

in the way a dry-construction wall is built. The pyramidal design is the same, and nothing needs to be changed except that fairly large stones should be used from top to bottom. When the wall is put up wet, (Fig. 22-3), drainage should be provided by leaving chinks in the lower portion or by inserting drainpipes through the wall. Dry construction is recommended here because of the thickness of the wall at the bottom, which makes continuous chinks or drainpipes difficult to fit.

The differences between a coursed rubble retaining wall and a regular one are also minor. A retaining wall, however, should be sloped backward into the earth. Don't use too many smaller stones, either, which can be dislodged easily. For larger walls, make the bottom thicker than the top. It doesn't make much difference for small walls.

Dry-wall construction is recommended for coursed walls as well, because the water will seep out through the chinks. If you do use mortar, it is easier in coursed construction to lay in drainpipes.

23

Using Ashlar Stone

Ashlar stone is quarried or factory-cut into pieces thinner than rubble (Fig. 23-1). Sizes can vary considerably, but the stones are usually one to six inches thick. You specify uniform thicknesses if you wish, but it is considerably cheaper to buy this type of stone by the ton and take potluck. As mentioned in Chapter 21, you can also buy stone in square-cut shapes and exact sizes, but this extravagance is only justified for indoor and formal patio use.

Unlike rubble, ashlar stone is well suited for use as veneer. Here, however, you *must* have a uniform thickness (usually four inches), since the stone is being laid face foreward. A variation in thickness would cause some pieces to project more than others.

STONE WALLS

Free-standing, ashlar-pattern walls are not as common as rubble walls. The pieces are smaller and more easily knocked down. For this reason, such walls are usually mortared. Construction is much like coursed rubble, except that the stone is thinner. To keep this kind of wall strong, coping and pilasters of block or brick are usually interspersed every six to eight feet.

Fig. 23-1. Retaining wall and stairwell show the versatility of ashlar building stone.

The most common application of ashlar stone is in retaining walls. The stone is usually put up dry. Drainage is much easier and construction goes much faster.

If you plan on building stairs, save some large pieces of stone for treads. When the stone arrives, however, don't be distressed if the truck driver dumps the load and smashes up most of it. He is saving you a lot of future work.

In some cases, a concrete footing will be necessary. Like all footings, this one should be placed six inches below the frost line and at least two feet below the surface. You should not need a footing with a mortarless wall unless the soil is very mushy.

When you build with mortar, the technique is much like bricklaying. Lay down a bed of mortar on the footing, butter one end of a stone piece, place it, and strike off the excess. Butter another piece, press it firmly against the first one, and so on.

You will run into trouble if you try to make even courses and neat mortar joints. Just put up a piece of stone with the straightest edge out, fill all joints with mortar and go on to the next. The only *bond* you should be concerned with is that of not laying one joint on top of the other. Try to use stone of the same thickness for each course and keep the tops generally level; you can adjust unevenness as you go along. Use varying-size mortar joints to take up the slack.

For dry construction, start one or two courses below the surface, making the joints tight and the tops level. Use extra-large, thick pieces of stone for this part because it is, in essence, your foundation. The soil should be dug out with a square-end garden spade to a depth of about four inches, and a bed of sand should be laid first (unless the soil is sandy anyway, in which case this step is unnecessary).

As long as you aren't using mortar, you can be even less fussy about fitting the pieces. Stone is very heavy and the surfaces are all essentially abrasive. When stone is laid on stone, the little projections lock against each other and are very difficult to dislodge. For that reason, try to get as much bearing surface between each surface as possible. But don't worry about gaps or uneven courses unless the stones tip and lose contact with the ones above and below.

Fig. 23-2. A dry ashlar retaining wall is made up of odd-shaped pieces, but the overall effect is a uniform one in front. Note the large pieces in the corner and the jagged pieces extending into the fill area behind the wall.

It doesn't matter what the sides and back of each piece look like, since no one sees them. Actually, the back should be as jagged as possible to provide better ties with the soil (Fig. 23-2).

Sooner or later, the stone broken by the truck driver will run out, and you'll have to start breaking or cutting some new ones. For a truly rustic look, just drop the stones on a hard surface and they will break in lots of pieces. You should be able to get a relatively straight surface on one of the edges and, if you don't, you can always cut the piece in half, thereby getting straight edges on two pieces. But try using the uneven pieces without cutting. You'll be surprised how straight the wall will look overall, even though individual pieces are rather jagged.

If you prefer straight edges, tap a line on the surface with your mash hammer and cold chisel. If you're looking for a high degree of accuracy, you should tap all around the stone, including the edges, but this is very time-consuming. Most of the time, you can get a pretty straight cut by scoring just the top. Once the top is scored, give the chisel some heavy whacks up and down the line, and eventually the whole line will crack and fall into two (or more) pieces.

For greater wall strength, taper your wall backward two inches for every foot in height. Also tie the wall into the soil behind it by inserting a long piece sideways every 24 inches. The best pieces for this are shaped like isosceles triangles.

Backfill when the wall reaches six inches in height. If your soil is highly organic, use a layer of sand directly behind the wall about six inches deep. Tamp down firmly and soak with water. When dry, tamp down again and fill with more water. That way, the dirt gets packed tightly, bonding with the jagged edges of the stones.

As you work through your stone pile, you should set aside thicker pieces with squared ends. These will be used in your topping course or stairwell, if one is planned. Some stone dealers advise mortaring this course even if the others are dry, but chances are that the mortar will crack with the frost anyway. You're better off simply butting the ends together and packing in the soil behind. Kids and dogs may knock these pieces off, but you can simply set them in place again.

STAIRS

An easy and economical way to make good-looking stone steps is to lay a veneer over existing concrete. If you like the rustic steps in Fig. 23-3, however, they must be built from the

Fig. 23-3. A complete retaining wall and stairwell can look attractive in spite of many irregular pieces used in construction.

ground up. They may be built separately or as part of a retaining wall.

Stone stairs should be built into the grade rather than projecting from it. Thus, the sides of the stairwell and risers are miniature retaining walls and should be constructed as such. The stairwell should be dug out in its entirety before starting and, if combined with a retaining wall as shown in Fig. 23-3, built along with the wall. Outside corners should be constructed of large, heavy, square or rectangular pieces, and inside corners should be dovetailed as with brick.

When your wall gets to the height of the first stair tread (seven to eight inches), build a small wall on either side, and another foundation wall underneath the front of the tread. The side walls should go back far enough so that you have a solid foundation for the tread on both sides. If you don't have enough large, squarish pieces for treads, you can buy them extra from the stone dealer.

WALKS AND PATIOS

Stone has long been a favorite for handsome patios, sidewalks, and other horizontal applications. *Flagstone* is nothing more than any hard stone that will split into pieces suitable for paving. Most of the stone mentioned will do this.

Again, it is easier and less expensive to lay this type of stone dry, and the project is accomplished in exactly the same manner as mortarless brick (Chapter 6). Here, of course, *bond* is meaningless, even with cut stone. If you prefer to use mortar, use mortar for the bed also instead of sand, and fill up all the chinks, too.

A common, and worthwhile, variation is laying stone over poured concrete. It is usually unfruitful to put down a new slab underneath the stone, but if the slab is already there, you can lay stone on top. Apply the stone as described under brickwork.

OTHER USES

Stone makes an exciting fireplace or house front, and there are countless other applications. Plans are given in Figs. 23-4, 23-5, 23-6, and 23-7 for several projects. It is easy to make up your own varieties too.

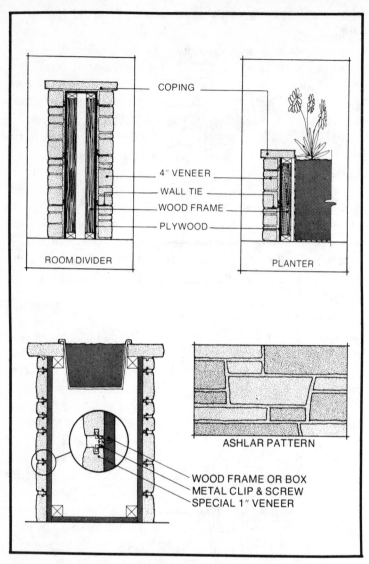

COPING

4″ VENEER

WALL TIE

WOOD FRAME

PLYWOOD

ROOM DIVIDER

PLANTER

ASHLAR PATTERN

WOOD FRAME OR BOX
METAL CLIP & SCREW
SPECIAL 1″ VENEER

Fig. 23-4. Room dividers and planters built inside the house can use both four-inch and one-inch stone veneers.

The methods of applying the stone are no different from the techniques previously described. Stone is used either in its natural position (as in walls or patios), or as a veneer. Follow the general instructions given above and adapt them to each specific application.

Fig. 23-5. An old-fashioned rock-slab outdoor fireplace.

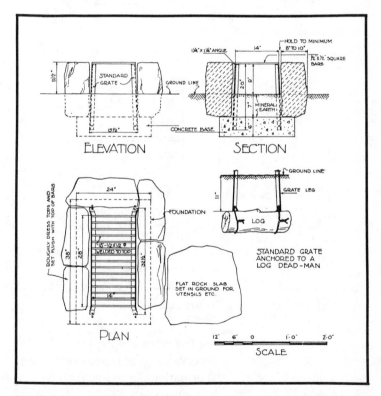

Fig. 23-6. A modern stone-faced barbecue with concrete backerblock and firebrick interior (Courtesy BSI).

313

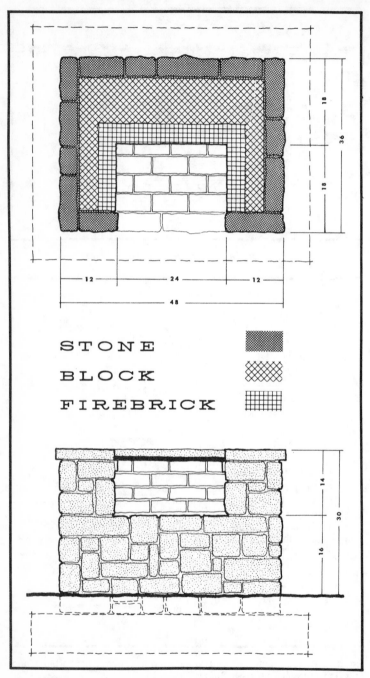

STONE

BLOCK

FIREBRICK

Fig. 23-7. Fireplaces can be constructed of stone, block, and firebrick.

314

Index

316